ROOT
RISE
ROAR

TRANSFORMING TRAUMA *into* *your* **BRAVE** *and* **BEAUTIFUL LIFE**

WRITTEN AND LIVED
BY
DAWN KING

Root Rise Roar: Transforming Trauma Into Your Brave and Beautiful Life
Copyright © 2023 by Dawn King

All rights reserved. No part of this publication may be reproduced, distributed, or transmitted in any form or by any means, including photocopying, recording, or other electronic or mechanical methods, without the prior written permission of the author, except in the case of brief quotations embodied in reviews and certain other non-commercial uses permitted by copyright law.

Printed in the United States of America
Hardcover ISBN: 978-1-958714-52-2
Paperback ISBN: 978-1-958714-53-9
EBook ISBN: 978-1-958714-54-6
Library of Congress Control Number: 2023933076

Muse Literary
3319 N. Cicero Avenue
Chicago IL 60641-9998

TABLE OF CONTENTS

Acknowledgments v
Introduction vii

ROOT 1
Chapter 1: Our True Nature: The Original Therapist 3
Chapter 2: Emotional Medicines 23
Chapter 3: Slow Down Your Crazy Child 63

RISE 87
Chapter 4: The Secret Life of the Soma: Singing 89
Chapter 5: Mindful of Movefulness 103

ROAR 121
Chapter 6: Courageous Communication 123
Chapter 7: Roar Reverence 141
Chapter 8: Growing into Congruency-Triggerless and
        Healed Trauma 159
Epilogue: Your Sacred Seat A Home Where You Are Safe
        and Warm 185
R3 Experience Root. Rise. Roar 187
Bibliography 189

ACKNOWLEDGMENTS

This book is a shared collaboration of teachings with and from the brave and beautiful clients who I am so grateful to work with. You are extraordinary in your capacity to connect to your courage, build mental strength, develop emotional fluency and create lasting change in your relationships, and in your nervous system. Your willingness to stay the course, peel off the layers, shed the skin and grow forward from the pain and the victory inspires me beyond the moon and back. You are the bravest people I know.

Thank you to my children. For your strength, compassion, forgiveness, love, family values and honesty. I am forever proud and awestruck for the remarkable individuals you are.

Thank you Leslie Stewart, Thomas, Brian Husband, Cora Boo, Sue Adamson, Mandy Schumaker, and Dr. Stevensen. If we are fortunate, a rare few people will stay and believe in us during the darkest of times. Emotional and traumatic pain has countless tactics to destructively protect itself. Understandably it can be very difficult for others to wait, accept or keep the faith in our restoration. Thank you for staying with me and creating formidable transformations that have occured in me because you stayed.

Thank you to my dear friend Corinna Stevenson. You have woven your tapestry of practices of nature interconnection into my life. You are a wayshower for so many of us. As a child my most influential guide and protector was nature, thank you for returning her home. To my supervisor Kristy Higgins. Your continuous belief in my work and my original way allows me to forge ahead. You are a pillar of strength and faith in the known and unknown.

Thank you Sara Connell for your powerful grace and trailblazing spirit. Thank you to the mighty MUSE Literary team for opening the door for the *Root, Rise, Roar* message and mission. To Mary Balice Nelligan and Megan Jackson, thank you for capturing, structuring and editing my creative windstorms into an orderly calm.

And you, dear reader and traveller, thank you for listening to or picking up this book. May its message ease and transform your pain and return you to the remarkable person you already are.

INTRODUCTION

On March 27, 1964, Alaska's mainland felt the magnitude of a 9.2 earthquake, which wobbled Seattle's Space Needle 1,200 miles away. The earthquake was so powerful it registered in Canada and every single state in the United States of America, with the exception of three.

That day was also Good Friday—the Earth trembled from the earthquake in a northern Canadian hospital, while the magnitude of contractions my mother experienced during my birth surpassed a 9.2 earthquake. I was born March 27, 1964, at 5:36 p.m.—the exact time the Alaskan earthquake hit. To this day, my mother jokes that the twenty-eight hours of physical labour and the trauma of my birth caused the earthquake.

As a child, I have no memory of safety, so my mind created safety in an imaginary world of song, animals, and nature. I learned and designed a hypervigilant radar screen, feeling what others felt. Instinctively, I seemed to possess x-ray vision for observing other people's behaviours, waiting and predicting what danger would come next. I was right every time.

My dad came to my bedside when I was five years old to say he would be leaving because my mom was an alcoholic. I did not know at the time that this was a common exit strategy for my dad. Previously married with children, he always had another woman to escape to; that was his relationship passport to emotional avoidance and escapism. By blaming the women or the alcohol, my dad always ran into the arms of anything or anyone that would justify his actions. This announcement terrified me, and I immediately experienced a "sensory unsafe" overload. Was my mom going to die? Where was I going to live? My roots were ripped out of the ground.

From that moment on and for the next 20 years, I could tell you when one ounce or one drink entered my mom's bloodstream. After all, by the time I was five years old, my vigilance to protect myself was mastered. My dad left me, my mom, my two older brothers, and our home, while she was in the hospital having a hysterectomy—a major surgery. My mom came out of the hospital and found the bank account empty and my dad with a new address and a new woman. Now alcohol could have its way with her, and it began to numb her emotions until she disappeared into a 20 year abyss.

After this event, my dad in his learned survivalist skills did an effective job of verbally and emotionally grooming me into the belief my mom was the cause of it all and he would save me from her. He promised the pony. I look back now and understand at every level why Mom continued to dive deeper into the shackles and chains of alcohol. She was injured by a partner who betrayed and pulverized her emotionally and mentally. My mom and dad were emotionally underdeveloped and unprepared for relationships based on the dysfunction of their own family of origin. Both of my parents were emotionally wounded, invisibly and profoundly. As I write this, my mom is in her eighties and has not had a drink in over thirty-five years. She still carries the debris of shame, guilt, and oppression by a generation which supported and normalized subservience. My father's dysfunction and destructive dependencies were in unhealthy relationships as well. Married seven times, repeating what his parents modelled until he passed away and sadly he died alone.

Looking back on my life, I see how my parents' traumas, thoughts, beliefs, and behaviours were transferred to me on very unconscious and seductively subtle levels. While genetics are undeniably inherited, I believe the learned behaviours of our parents or caregivers and environments as children have a constructive and destructive effect on us. We can also change them. My mother was adopted, and she gave up finding her birth parents. She has been on depression medication for most of her life and married three times, historically to abusive or emotionally unavailable partners.

My father had a childhood and upbringing that was void of any stability, emotionally or physically. He wore a "man's man" strong, charismatic, rigid armour on the outside yet was also a traumatized, fearful and regretful man on the inside. He never took those protective shields off and was burdened with unexpressed anger, grief and regret.

I stayed in the home that generations of hell built, until Mom and Dad followed the dopamine of chemical dependency and moreaholic relationships. At fourteen years old, my mom walked out of our one bedroom apartment and my father never did come back from searching for love and attachment in all the wrong places. The survivalist in me knew it was time for me to quit school. At fourteen years old I washed hair at a salon during the day and cooked at a local golf course at night. I took over paying the rent and all responsibilities of an accelerated adult life.

Many of us—in fact, most of us—have experienced traumas in numerous ways. Trauma responses are a separation in us, while recovery our nervous system happens when we enter a new relationship with it and others. The traumatic pain appears to have guided me in a way that actually helped me. I had to return to the past to find my roots again. Our childhood, or deeply painful life experiences never leave us, but the avoidance of our emotional pain throws us into either numbing or escaping it in some way. I thought by writing this book I could share with you how I looked at my past, felt the pain, and how it transformed me. The following pages convey what trauma taught me. I felt compelled to share what I learned from adversity, trauma, and loss, and how we can move through it.

I think we dismiss and avoid pain, instead of visiting and remembering its wisdom. Through defences primarily learned in childhood or enhanced as we grow older, we deny, think positive, forget about it, inflict self-harm, or medicate the discomfort. All the while, it simply wants us to slow down and hear its voice.

Trauma is inherent in all our storied pasts. We are stuck and imprisoned in the unheard, untold stories of our pasts. It is here that the reconciliation of our individual stories can find relief or liberation. But it is here, in the roots of our history, that we can also find the resilience and courage to take that pain and transform it into a brave and beautiful life.

Physical or emotional traumas can be born in different external environments. But the exact safety-seeking survival responses of "not feeling" are identical. Had I known how generational traumas and the beliefs or behaviors of my parents would impact my future, I am certain I would have likely stayed nestled in the womb.

More earthquakes and tsunamis have hit the shores of my life, washing over or drowning me since my mother's traumatic birthing experience. Through all of them, I have held on, sometimes barely, and other times ferociously, fighting for my life.

Traumatic experiences and responses are a part of every one of our lives, and oftentimes, they present as "mental health" issues. However, I like to think of these responses not as mental health problems but as the key to addressing somatic and psychological symptoms impacting the brain, body, and emotions.

Pathologizing, overmedicating, or mindset coaching our emotional pain act as Band-Aids to deep wounds deserving greater care. Suppressing our traumas or invisible wounds by thinking positively or avoiding them is not solving the problem. We have entered the age of anxiety, addiction, and overachieving where more of everything is better—eat more, medicate more, drink more, think more, exercise more, make more money, and basically, do more of anything.

In a world of hyper-stimulated, over-achieving, transformational development, we avoid our fear or difficult emotions by doing something all the time. We resist slowing down. We are emotionally guarded soldiers thickly armoured with overthinking, perfectionism, low self-worth cycles, people pleasing, addiction, misunderstood anxiety, and rotations of delusional denial.

We misuse or overindulge in alcohol, drugs, sex, relationships, exercising, fad diets, social media, vaping, phones, food, or overachieving. Yet, at the same time, most of us are "under-being." We are comparing our lives to others instead of learning how to connect what our own north star may be or where our compass wants us to go. In a performance-based, self-worth culture, overdoing can often create a disconnected journey at a Mach 5 pace as we go so fast that we may not hear key messages and our inner voice. Survival speed is rarely slow. We often feel we are chasing life or catching up to unrealistic expectations and productivity.

Happiness is for sale, and we buy it in the billions. We may be attracted to self-medication so we can simply escape our lives of chaos, although we really long for peace. The age of anxiety is here while the delusion of comfort has seduced great people with superficial care and strategies. The workaholic world is normalized, and the bypassing of pain is big money. The comfort zone may be killing us slowly or quickly as temporary relief for the pleasure-seeking brain, and so we repeat.

As a registered counselor, Canadian-certified addictions counselor, and trauma therapist, I specialize in helping others with their relationship with themselves and others. I help others heal from the effects of various types of traumas by helping them find their voice, overcome their fears, and get honest with all of the avoidance. As a result, a life of worth and purpose can root and then rise.

Every day I see people shackled with guilt, shame, denial, and stigma. Our minds and emotions live in an external world.

I felt compelled to write this book in the hopes I could reach more people and help simplify the process of recovering or restoring from anything. The practices and principles of this book are all learnable. I discovered and created them because

I could not find the help I needed. I started to look to the past, not the future. I went back to my roots and found what didn't work. I was not aware my life had been my experiment to heal and redefine. This may be true for all of us. Trauma created numerous strengths in me, and for over thirty years I have learned, unlearned, taught, and lived them each day. I was diagnosed and have fully recovered from substance use disorder, post-traumatic stress disorder (C-PTSD), and major depressive disorder. Before all of this, I was obsessed with self-help books and seminars. I thought two marriages would make me feel safe. Eventually, I drank my pain, trauma, and grief into the psychiatric ward. I also checked myself into addiction treatment twice and tried to meditate, change my mindset, and medicate myself into calmness. I have been seeking pain relief for most of my life. I started at a support group for Adult Children of Alcoholics at fifteen years old, and I carried on curiously and openly seeking help for over forty years. While I am a supporter of whatever works for an individual, the hundreds of ways I tried to find my way all lasted, temporarily. Although temporary, each step of the way there was always a knowing and unseen sense that I was finding my way. Each failure and every step were patterns showing me the feedback and progress. There is no right way to do life, it will have its way with you. As long as there is breath, keep going.

I've been deeply saddened by the lack of effective care and collaboration between alternative and traditional healthcare. I've also felt anger and frustration at the lack of effective and progressive long-term care. Yes, crisis care, and quick fixes are readily available. It is my hope we create more mentors and teachings which demonstrate how to heal or recover from emotionally destabilizing. We can then continue to move people through the sustainable stages of long-term recovery from anything. Many well intended individuals and professionals are putting the fires out, applying the band aids or offering the quick fixes for our mental or emotional support systems. But the gaps of childhood or behavioural development keep people falling through the cracks. Many of us simply need a guide map towards growing up and into a healthy adulthood.. Through my journey, I've learned the importance of bridging allopathic and natural healthcare, drawing upon strengths from both. Ancient wisdom and modern technology are equally profound. My pain and need for lasting change drove me to dig deeper and create solutions.

With or without this book, I hope you find your way. The practices and principles of *Root, Rise, and Roar* has kept me and many others fighting, growing, feeling, and finding our way.

Most of us have walked on splintered bridges, using defective survival techniques which have been ineffective in helping us move from childhood into adulthood. Our family and environments or our rights of passage were mostly dysfunctional, distorted, and unsupported. We carry these experiences with us for the rest of our lives until the pain asks us to change. Do these experiences have to create needless suffering? No. But do they require care and understanding? Absolutely. It is my hope this book serves as a guide to reduce your suffering, heal your past, and help you rise into a life that is not rid of discomfort, but one in which you find your voice and congruently roar like hell while being rooted in your real and courageous self.

I also hope this book serves as support to other therapists, coaches, and trauma-informed professionals. Perhaps after reading these pages, you may understand your clients and serve them or yourself, even more. Whoever you are, dear traveler, thank you for doing such brave and beautiful work.

Emotional pain and trauma are powerful teachers and extraordinary guides. I hope that these pages and practices transform any parts of your pain or trauma into beauty and bravery. When we heal, recover, change and become accountable for our pain we can become way showers and generational cycle breakers. When we heal it, feel it, own it and transform ourselves each generation can be freer and healthier than the one before.

ROOT

"The sorrow, grief, and rage you feel is a measure of your humanity and your evolutionary maturity. As your heart breaks open there will be room for the world to heal."

—*Joanna Macy Ph.D.,*
The Work That Reconnects

CHAPTER ONE

OUR TRUE NATURE: THE ORIGINAL THERAPIST

Feral Child

The forest, the tundra, the waters, the skies.
A sanctuary of safety, for a little girl in disguise.

Tiptoeing as a whitetail, her womb is the wild.
Humans have hunted and harmed her; she is a feral child.

She evacuates her body, to swing from the stars.
This a resting place, to bandage her wounds, and caress her scars.

Falling, she whistles to the northern lights, the lights hear her call.
Never one to leave a wounded child, Aurora embraces her fall.

Surfing the solar waves, nestled in Aurora's girth.
Reluctantly, the little one floats back to Earth.

Hidden in the blackberries, protected by their thorns.
Her breath is strangled repeatedly, fear is reborn.

Humans returning to hunt, she was stalked and taken.
It is time to leave her body, where her spirit will awaken.

Reconnecting to her body, lifeless and battered.
With a moist breath, a nudging nose, the pack has gathered.

She has returned home.
To her tundra, the muskeg and the flocks.

Here is home and healing.
The little girl the Wild never forgot.

—*Written and Lived by Dawn King* ©

* * *

Mother Nature is and remains the original therapist. I received therapy from Mother Nature as a child, and she remains one of the most potent healers.

As a four-year-old, I had conversations with my cat, the bees, and blackberries in my backyard in Vancouver, British Columbia. These connections brought me a sense of calm in my chaotic, abusive home. Nature listened to me with a skill and presence that adults in my life could not resolve.

When I was afraid of the dark, I spoke to it. I remember listening to the owls in the black of night as they hooted their presence with a message that still stays with me decades later—if an owl can see in the darkness of life, so can I.

As I grew older, I listened to the northern lights. I also heard the sounds of snow, captured the whispers of the winds, and sang with the birds whenever I was lonely. Nature was a nurturing presence, a family of friendly voices, soothing textures, vibrant sounds, and vast languages I could translate… It eased my pain, gave me an escape, and was something to enjoy—a nourishing connection and undeniable communication. Animals and nature created the connection I did not have with humans. I felt a sense of belonging with nature—the home I never had. Nature had a way of understanding me. She just listened and showed me the way.

Many children of abandonment, abuse, or neglect relate to and find a sense of belonging in nature. Our psychological responses to trauma—our survival skills—send us seeking safety and in search of emotional poultices. This relief from nature can generate feelings that soothe and calm or invite feelings of wonder and joy. In nature, we find kinship and interconnection.

Early connections to animals, plants, trees, clouds, nature sounds, insects, winds, flowers, and any other elements of nature provide nurturing and can repeatedly be found in the stories of adults who experienced adverse childhoods. One of my clients leads a multinational company. He has social anxiety, and after spending time with me and a few months of exploring nature, he credits the winds in the willow trees in his backyard with soothing his performance anxiety. In the "always do more" and "enough is never enough" culture of sales and productivity, my client learned to relax in his mind and body by listening to the winds, then embodying the sounds and sensations in his imagination, which eventually taught his entire nervous system to slow down. His internal world was back in control of the external push and pressure of his job.

American media host, producer, author, and actress Oprah Winfrey shared in a recent interview that walking her five dogs in nature is how she starts her day. As the internationally bestselling author and researcher Richard Louve puts it, "When we don't spend time in nature, we develop nature deficit disorder."

Nature creates calming feelings for Canadian singer Justin Bieber, who still likes to make a campfire and enjoy the serenity of the outdoors. During a summer outdoor trip in recent years, Bieber tweeted, "Enjoying time with my family literally in the middle of nowhere, but this nowhere place is calming and is pretty beautiful." On January 4, 1937, German-born theoretical physicist Albert Einstein said to his son, Hans Albert, "Look deep, deep into nature, and then you will understand everything better."

Recently, I had medical imaging taken of my knee. The MRI machine was loud and confining. Although I was diagnosed by a psychiatrist with PTSD, it had been many years, and rarely do I have symptoms of claustrophobia, racing heart, or hyperalertness brought on by fear of confinement. But I have to listen and be aware of my inner and outer environments to make sure I know how to calm and ground overwhelming sensations. This is a skill we can all learn for various symptoms of fear and anxiety. Our brain is created to keep us safe through alerted physiological responses, learning how to calm those responses is achievable but we may have never been taught how.

As the nurse gave me my emergency button to hold on to and ear plugs to put in, she asked, "Are you ready?" I looked up at the ceiling and to my mental and emotional relief, I saw a mural of clouds, trees, and birds on the ceiling. I took a deep breath and said, "Now I am, thank you."

Many clients have been referred to me with a common fear or anxiousness about going to a dentist. Becoming slightly nervous can be a normal feeling for many, but this is different. Fear can cause an avoidance to go to the dentist (or anyone or anything), whereas a phobia can cause extreme avoidance, extreme fear, and anxious distress. We may associate to past adverse experiences at the dentist, or we may feel confined in the chair or when someone stands over us. We may fear the noises of the machines we hear at the dentist. We can also fear what the dentist will find and futurize or catastrophize about the unknown.

Well-intended healthcare professionals are often not even aware they need to be informed of past traumas. That is, they may not even know to ask you for permission to start procedures, let alone verbally, emotionally, and psychologically prepare you for the process. This is where fears from the past or in our imaginations are created, consciously or unconsciously.

In my work, I begin to help people notice how they feel and ask them to describe those feelings in their bodies. Sensations such as tightness or tension are common. I ask them to describe how their breathing feels, and they usually express words such as shallow, short, or pressured. This could be due to previous dental office experiences or simply the confinement and sounds which may stimulate traumatic responses. Fear can be a minimized emotion or ridiculed or criticized with responses such as: "It's all in your head," "Man up," "Don't be such a suck," "Get over it," or "It's no big deal." We often tell ourselves these messages and hear them from others in conjunction with most fears.

First and foremost, when impacted by trauma, one needs to tell the doctor and staff how they are feeling. With a trauma-informed staff, they will hopefully slow down the entire process and patiently let the patient's body and brain adjust to these responses. Hopefully, forcing anyone through fear or phobias is a practice of the past. Letting us adjust and create new associations that teach our brain, body, and emotions that we can work and move through discomfort is most effective. Avoiding painful situations is not the answer to working through fear or the fight/flight response. Slowly and consistently, we can learn to adjust or synchronize overwhelming thoughts and senses. This is a process that requires patience, and no one should be forced into anything.

The body keeps score, as Dr. Bessel Van Der Kolk writes in his book of the same name, and this opens up the possibility for the person to learn a new experience. Dr. Van Der Kolk describes it as, "befriending the body".

> *"Trauma victims cannot recover until they become familiar with and befriend the sensations in their bodies. Being frightened means that you live in a body that is always on guard. Angry people live in angry bodies. The bodies of child-abuse victims are tense and defensive until they find a way to relax and feel safe. To change, people need to become aware of their sensations and the way that their bodies interact with the world around them. Physical self-awareness is the first step in releasing the tyranny of the past.*
>
> *All too often, however, people self-medicate, or drugs are prescribed instead of teaching people the skills to deal with such distressing physical reactions. Medications can work but they can also blunt sensations and do*

nothing to resolve them or transform them from toxic agents into allies.

The mind needs to be re-educated to feel physical sensations, and the body needs to be helped to tolerate and enjoy the comforts of touch. Individuals who lack emotional awareness are able, with practice, to connect their physical sensations to psychological events. Then they can slowly reconnect with themselves."[1]

We learn the benefits of disassociation very early and often in childhood. Most of us who have experienced physical, developmental, or psychological traumatic events have all done this. It is a primal strategy for our brain to protect us. Trauma responses are the bodily sensations, thoughts, or feelings that do not return to their original state before the event or continuous stressful environments. We can usually all remember a time when we "checked out" emotionally and mentally by disassociating from the situation with thoughts, images, and emotions. Fantasy and imagination are powerful medicines in dissociation. They can be highly effective ways and saviors when the nervous system and brain simply cannot tolerate the load.

The songs, sounds, and connections I felt from nature soothed a highly alerted electrical system within my brain and body. Using my imagination and talking to animals, trees, insects, and the earth calmed the survival part of my brain and gave me an inner connection to myself.

---

[1] Bessel A. Van Der Kolk, *The Body Keeps the Score: Brain, Mind, and Body in the Healing of Trauma*

As a ten-year-old living in the Northwest Territories of Canada, the tundra became my emotionally dissociative equalizer and dear friend. I would escape into the night, walking by myself in the muskeg and tundra. The frozen flatlands were whipped by the arctic winds, and I was mesmerized by the swirling halos of northern lights. I felt alive and settled within when I talked to them. Alone, in the middle of the night, I would lay against the ice walls of the Great Slave Lake where I could heal the physical wounds of the beatings.

The cold winds of the North were my anti-inflammatories, the howling wolves sang to my sorrow, and the winds were my salve. Nature became my healer, my little girl's safety and sanctuary from abandonment, abuse, and adult addictions.

While we all have the ability to dissociate, our brains will often shut down the traumatic event by not remembering—yet it stays in the nervous system. Sometimes we can have a feeling of constantly being on guard. When the nervous system has a disruption (which is unique to all of us), such as childhood trauma, accident, chronic illness, abandonment, and other life events that leave us feeling unresolved loss and fear, our nervous system doesn't immediately settle down. It's going to be turned up for a while, alert for the possibility of further danger.

For example, you might keep looking over your shoulder or be constantly scanning your surroundings for threats. You've been hurt before, and you don't want to be caught off guard. It really means your brain is doing its job to protect you, although this knowledge doesn't make it any more comfortable to feel on edge all the time.

Another way unresolved trauma can appear is through being easily startled. This is an indicator the nervous system is temporarily stuck in the "high" setting, and it is going to be easily startled by things like when the phone or doorbell rings, a door is accidentally slammed, or a person unexpectedly walks into a room. Many of us are claustrophobic or feel confined in specific environments or situations. These physical, emotional, and behavioural responses are the nervous system's way of sharing the past with our bodies and brains. You may find yourself jumpier than usual. An example of not consciously remembering can appear with difficulty sleeping. Sleep is a vulnerable state, and when the brain and body are revved up, we're likely to have a hard time sleeping. It's as though the mind is saying, "Danger! This is no time for sleeping!"

Trauma is not "just in your head" or relieved by just "changing your mindset." Traumatic events can stay hidden until something triggers us. Chronic stress, the loss of a job or loved one, financial stress, prolonged illness, going to a dentist, or any specific situation or action that leads to feelings of chronic worry or ruminative fear are known as anxiety builders and triggers. Triggers are sensory stimulations connected with a person's trauma, and dissociation is an overload response. Even years after the traumatic event or circumstances are no longer present, certain sights, sounds, smells, touches, and even tastes can set off, or trigger, a cascade of unwanted memories and feelings.

When triggered by a traumatic event, we may experience sudden urges of panic, depression, or impulsive anger. Traumatic events push the nervous system to restrict its ability to regulate itself. For some, the system gets stuck in the "alert" position, and the person is overstimulated and unable to calm. Anxiety, anger, restlessness, panic, and hyperactivity can all result when you stay in a ready-to-react state.

The body's "alert" state, called the "fight or flight" response, is controlled by the sympathetic nervous system, which is part of the autonomic nervous system. The other part is the parasympathetic nervous system, which works to relax and slow down the body's response to stress. The sympathetic and parasympathetic nervous systems act like the accelerator and brakes on a car. The sympathetic system is the accelerator, always ready to rev up and take us out of danger. The parasympathetic system serves as the brakes, slowing us down when danger isn't present.

Nature has a natural way of having a calming effect through healthy dissociation. We can connect emotionally to animals, sensorially attach to trees, winds, and sounds, distract fearful thoughts, and leave the environments of traumatic response through our parasympathetic nervous systems. The parasympathetic nervous system controls the body's ability to relax. It's sometimes called the "rest and digest" state.

The "do-aholics" or fast-paced culture of our society and in our homes today has the human body constantly responding to non-life-threatening stressors with the same fight or flight response that can cause high levels of anxiety. Chaos and busyness are normalized. In my country of Canada, doctors are now legally able to prescribe free passes to our provincial or national parks as a "prescription" for mental wellness.

Research has shown that the long-term effects of chronic stress affect a person's psychological and physical health. According to an article in *Harvard Health*, "The repeated activation of the stress response takes a toll on the body. Research suggests that chronic stress contributes to high blood pressure, promotes the formation of artery-clogging deposits, and causes brain changes that may contribute to anxiety, depression, and addiction."

The changes in the body take place very quickly when the sympathetic nervous system is activated. Until the brain perceives that the danger has passed, it continues to release hormones that keep the body on high alert and ready for intense physical activity. Once the threat is over, cortisol levels decline and the parasympathetic nervous system slows the stress response by releasing hormones that relax the mind and body while inhibiting, or slowing, many of the pre-existing high-energy functions of the body such as rapid breathing, increased heart rate, etc.

When the parasympathetic nervous system is activated, it produces a calm and relaxed feeling in the mind and body. People can learn to trigger their own parasympathetic nervous system. This is where nature is a profound healer. By simply connecting with a natural environment such as sitting on the grass, holding a leaf, or looking at a tree or plant, someone can immediately reduce their sense of anxiety and stress. This also lifts their mood, strengthens their immune system, and reduces their blood pressure.

However, one does not have to drive to the forest or countryside or hike a mountain to receive the benefits from nature. Nature is everywhere, and it can be accessed easily. In my office, I have a plant… one plant, an African violet. A low-maintenance, easy-to-grow houseplant, these vibrant purple-colored, clustered flowers are attached to dark green leaves and succulent stems. This violet has been a remarkable and effective calmer and collaborator over the years

Most people I see have various degrees of anxiety. In my second-floor office building, I cannot walk out into nature with them, but I can introduce them to the plant. I ask clients who come in nervous, anxious, fearful, or experiencing panic attacks to begin breathing and focusing on the plant, describing what colors they notice. I ask them to touch the plant, feeling its textures. I ask them to describe the dirt the plant is in and to touch, hold, and feel the dirt if they wish.

A client who teaches at the university came to see me because she could not sleep and slow her thoughts down. Once she felt the plant's leaf, she instantly started to take a deep breath and sighed numerous times. This simple, powerful, slowing-down process moved her from sighs to tears. She had recently been divorced. The textures of the plant reminded her of her garden. Through the divorce her husband declared bankruptcy, and she was unable to hold onto her house. She had developed insomnia, anxiety, and depression, and her doctor was kind enough to refer her to me. As this client touched the plant's leaf, she began to share how sad she was not having her home and garden anymore and that she missed her family being gathered in the home, as they had done for twenty years. She then began to feel her sadness, apologizing for crying and expressing how horrible she felt about herself because everyone kept telling her to "move on" and "get over it." I shared with her that depression can often live in the past, while anxiety can live in the future…that often our untold stories of pain and sorrow are trapped, unexpressed, and our emotions are not given a voice. Our pain is as important as our pleasure. I shared with her that her courageous emotional vulnerability was welcomed here. The plant literally returned her to the roots of her past, and she grieved and relived in her mind, body, and emotions the stored hurt and loss.

There is a misunderstanding that clinical counseling only focuses on traumatic issues. I can tell you that one of the most powerful therapeutic effects we can give to another person is connection and active listening. Most of us really have not experienced relationships that listen to our feelings or fears without advice giving or fixing. Most of us are told to change our stories instead of sharing them. Nature has her ear to the ground, always listening and accepting all of our parts. That one plant in my office has been the rooted ground that allows so many people to release emotional pain and see themselves growing deep and healthy new roots.

I have many options to be with nature in an office building. A raven and a hawk visit the electrical pole outside my window daily. I also have numerous pigeons who perch on my window ledge to whom I regularly introduce to my clients.

A high-achieving business owner in his late thirties was referred to me for destructively and verbally expressing anger by yelling and demanding people get things done at his workplace. He came into my office, resistant and guarded. Anger often stays shackled in the body this way. In his mind, there was no way I was going to help him. He didn't want help. He was here because his employer made him come, and he was going to lose his job if he didn't.

After three sessions of testing my acceptance of his stoic, defensive demeanor, I continued to lean into his resistance, gently listening and asking him about his history, starting from childhood. Why a pigeon arrived on my windowsill during this man's session and began to annoy him with its intense and relentless cooing, I will never know. But the pigeon, combined with history, opened the feelings door to this man as he talked about painful memories the anger had guarded. This annoying pigeon triggered and stimulated frustration in this client repeatedly. The man began to notice he blamed the pigeon for his anger just as he did with coworkers. Over time, this highly intense achiever of a man shared his traumatic stories with me and the pigeon. As we both annoyed him and accepted his stories simultaneously, he began to express emotional pain. Anger was taught to him by his father as a display of strength and power. He learned that anger could control and rigidly protect his vulnerable feelings.

Hurt, betrayal, loneliness, sadness, and grief can all be armored by anger. It protects the discomfort of feeling these emotions; imprisoning them will not allow remembering or expressing them in healthy ways. Most of us are not our destructive behaviours—we are primarily our unresolved emotional pain. This pigeon was as feisty and as vocally assertive as my client. By session four, my client's physiology began to soften. He started to breathe, to calm, and to settle. The tension in his body unwound. Many months later, through the relief of sharing his feelings about his traumatic past, he walked into my office on a transformative day, went directly to my office window, and asked, "Have you seen the pigeon this morning?"

In addition to random acts of pigeons, I keep numerous shapes and sizes of feathers in my office. I have clients who have learned from our session time how to use nature to calm themselves. One client, immediately upon entering the office for a session, will sit right at the window looking for the birds to arrive or pick up a feather before she begins her session.

I have numerous shapes and sizes of rocks in my office as well. A successful entrepreneur came to see me because she was diagnosed with bipolar and major depressive disorder, and she always gravitated toward one particular rock. This client expressed to me how soothed and calmed she begins to feel when touching and seeing the rock. After a few months together, she now has numerous calming antidotes in her repertoire, including rocks, feathers, snowshoeing, hawk watching, and her cats.

A corporate man in his fifties sees me for recovery from workaholism. He enjoys picking up, holding, and rubbing a rock that is on my table. He notices his jumping leg and inability to concentrate his thoughts or focus changes right away when he does this. He tells me the rubbing of the rock gives him relief and "slows the inside nervousness." The rock's smooth edges and the soft cool textures he says, "Brings him back to a childhood babbling brook." He used to play in and sat on its shores for hours under the coolness of the coastal giant cedar trees. His body and brain remember how the river rock pacified and alleviated a harsh and verbally abusive home environment. He picks up this rock before each session as it grounds and calms him. The rock returns him to emotional stabilization. Eventually, through our sessions, he learned the root cause of his anxiety and fears, and he found his own rock to keep with him in his pocket as a way to remind him to breathe and regulate physiologically.

Often triggers refer to emotional disruptions we do not want. But this rock triggers the childhood sensories and emotional sanctuary this brook provided him in his past. Remembering through the rock, he has taught his parasympathetic nervous system to ground and slow down, feeling safe again.

Feeling a feather or focusing our eyes on a hawk, raven, pigeon, cloud, leaf, tree, rock, or plant within my office provides therapy for which I take absolutely no credit. I could never be that effective. Nature is the original therapist, and she is our constant companion in connecting our mind, body, emotions, and spirit and in relieving the anxiety, tension, and fearful thoughts of our past or imagined experiences. Nature graciously accepts all of our parts and creates a container in which we can self-regulate. Nature doesn't have to learn how to self-regulate; she already does. This is exactly why she is a profound teacher and healer.

All of us experience various degrees of trauma, and psychologists describe them in three main categories:

- Acute trauma results from a single incident.
- Chronic trauma, such as domestic violence or abuse is repeated and prolonged.
- Complex trauma is exposure to varied and multiple traumatic events, often of an invasive, interpersonal nature. These include:
- Accidents, bullying/cyberbullying, dysfunction at home (such as domestic violence, parents with a mental illness, substance abuse or incarceration), the death of a loved one, emotional abuse or neglect, physical abuse or neglect, community violence (shooting, mugging, burglary, assault, bullying), sexual abuse, natural disasters such as a hurricane, flood, fire, or earthquake, or war and acts of violence.

What is unique about trauma responses is that we all respond differently to events, environments, or situations, but the effects of nature on calming and connecting our brains and bodies is as old as time. What we can all share or experience is the innate ability nature has to inherently regulate our wildly reactive or ruminative thoughts or associations to the events of the past or the future.

I wrote the poem at the beginning of this chapter many years ago. The feral child was me. When the feral child is returned to the hunt (various degrees and forms of abuse or abandonment), dissociation serves as a protection to sense imminent danger. But, ultimately, it was nature who protected and preserved me.

My body, mind, and emotions learned that everything was dangerous or certainly worthy of paying attention to in an alert or hyper-aroused state. With eyes and ears open, always scanning, I embodied the role of "danger detective".

We have been raised for generations never having learned to talk openly about our feelings. We talk more openly and tell stories of our physically broken bones and sprained ankles than about the emotional pendulums real life delivers. Learning to cope with emotions is where great work is done, yet nature does this effortlessly.

We are a fast-paced "do-aholic" and "more-aholic" culture that rewards more of anything, distracts from what is uncomfortable, and self-medicates for temporary pleasure or relief. Nature still remains the ultimate therapist, as she is self-regulating.

"There is mounting evidence, from dozens and dozens of researchers, that nature has benefits for both physical and psychological human well-being," says Lisa Nisbet, Ph.D.—

a psychologist at Trent University in Ontario, Canada, who studies connectedness to nature. "You can boost your mood just by walking in nature, even in urban nature. And the sense of connection you have with the natural world seems to contribute to happiness, even when you're not physically immersed in nature."

Increased scientific support documents time in nature can act as a balm for our busy brains. University of Chicago psychologist Marc Berman, Ph.D., and his student, Kathryn Schertz, explored this in a 2019 review. They reported, for instance, that green spaces near schools promote cognitive development in children, and green views near children's homes promote self-control behaviours. Adults assigned to public housing units in neighborhoods with more green space showed better attentional functioning than those assigned to units with less access to natural environments. Experiments have found that being exposed to natural environments improves working memory, cognitive flexibility, and attentional control, while exposure to urban environments is linked to attention deficits.[2]

Whatever you call it, connectedness to nature seems to benefit mood and mental health. In a meta-analysis, Alison Pritchard, Ph.D., ABPP, and colleagues with the University of Derby in England, found that people who feel more connected to nature have greater eudaimonic well-being—a type of contentment that goes beyond just feeling good and includes having meaningful purpose in life.[3]

Nature takes nothing from us. She adjusts and adapts within her own cycles and systems. Nature is always present, and in nanoseconds her companionship can be felt with a glance to

---

[2] *Current Directions in Psychological Science*, Vol. 28, No. 5, 2019
[3] *Journal of Happiness Studies*, online first publication, 2019.

the sky or the ground. She is everywhere and rarely in a hurry to get a result. We can be obsessed with outcomes, black and white or goal-oriented at all costs. Our obsessive, performance-based perfectionism life gets super simple in nature and can profoundly root us into remembering our real selves. Nature can remind us that we are not in control of everything and everyone. We are co-creating in a vast and magical ecology and trusting in the process while putting our best foot forward. Often it all turns out even better or differently than we had expected.

I urge you to try something I share with my clients. Go lay on nature's ground or the grass, stand under a tall tree, get your hands in the dirt, lay in a stream, swim in a lake or the ocean, reach out to leaves blowing in the winds, hold a cat, or simply touch a live indoor or outdoor plant, and the process will likely calm or ground you back into your body and maybe even slow ruminative thinking or help your brain to "shut off".

Nature teaches connection to self. She can slow us down as well as add emotional exhilaration, playing or exercising in her wondrous options: surfing, swimming, sledding, snowshoeing, hiking, sitting under the shade of the tree, swaying in a hammock, gardening, growing herbs indoors, listening to songs of birds, and even observing a persistent pigeon on a city office ledge. Even rubbing a rock can bring us back to the roots of ourselves.

While I am taking intermittent breaks from the screen and the mental process it takes for me to type words (it is way more natural for me to speak to them), I am thankful for my cat. All I have to do is reach over and rub under his chin, look into his eternally wise eyes, and we both feel calm and connected. We both begin to purr.

CHAPTER TWO

EMOTIONAL MEDICINES

"Perhaps everything terrible is in its deepest being something that needs our love."

—Rainer Maria Rilke, poet and novelist

In the Womb of My Own Care

I am in the womb of my own care. Silent and solitary.

I nest deeply in hibernation, in despair, in loss, in grief, a necessary home.

Restoration begins in the shadows of pain. The shattered pieces, the rigid dark edges, the jagged little lies of denial, the delusional views of safety, the cuts releasing the emotional pressure.

In the womb of my own care is where the wounds begin to reveal and to heal.

Excavating the riverbeds of tears, turbulent stirring of debris creates cleansing currents for the stagnant, suctioned emotions.

In the womb of my own care, I am an infant, I am a parent, I am a wise woman, I am a wounded child. I am a bridge for it all.

Nature has taught me to winter, to rest. The light of spring will find me, it always does.

I will awaken yet again, renew, rejoice and I will return to winter.

In the womb of my own care, I circle as an animal would paw and revolve into a bed of rest.

Nourishing and dreaming the shadows into light, I return to my emotional haven and nestle into winter home. Spring will find me.

—Written and Lived by Dawn King ©

* * *

I quit school in grade seven, lived on my own, worked two jobs, and lied about my age so I could get into barber school. I finished barber school at fifteen years old and with nature always being my true and original guide, I gravitated toward becoming a deckhand on British Columbia's Wild West coast. I discovered a very unique opportunity quite quickly—fishermen need haircuts. An entrepreneur was born, and there I was, a sixteen-year-old woman in a man's world. Forty years ago, there were very few women on the docks of the Canadian west coast, and those docks helped an independent, self-reliant young girl take care of basic survival needs such as food, water, and shelter.

There was no room for emotional vulnerability either. Feelings of fear or uncertainty had to be internalized and never shown. Confidence and courage had to create walls of protection which served as defence from potential dangers. These were all emotional resiliencies which constructively guided me to assess potentially dangerous environments. My feral child was turning into a teenager with situations and environments that called upon those skills of protection. My teenager's hypervigilance held hands with my feral inner child and quickly taught me how to read people emotionally, connecting those messages to body language with laser precision and accuracy.

Emotions are siblings.

A leading expert on violent behaviour, security specialist, and author of *The Gift of Fear*, Gavin de Becker explains this as a positive reaction when fear brings such things to our awareness like when a date won't take "no" for an answer, a new nanny gives a mother uneasy feelings, or a stranger in a deserted parking lot offers unsolicited help. The threat of violence surrounds us every day, but we can protect ourselves by learning to trust and act on our gut instincts, feel our emotions, and then watch others. True fear is a gift, unwarranted fear is a curse; learn how to tell the difference. This is a fabulous book, but I simply learned this concept naturally as a child while constantly in potentially dangerous environments.

Scanning potential threats to myself and my environment became my superpower for safety. I learned how to protect myself physically, and I figured out exactly how to do that by emotionally regulating first.

I am a person who passionately appreciates finding patterns in behaviour. Reflecting on and repairing the emotional development of my childhood has been a lifetime journey of determining and recovering from those patterns. I discovered there is a pattern that repeats itself with mostly all children who are raised with these gaps or blanks caused by trauma or stressful situations, and it carries an obvious message, "Don't feel." The avoidance of feelings is costly, for many reasons.

We learn through our families, caregivers, teachers, and society not only not to feel but also that when we do feel strong emotions, we are to avoid them or block them all together. As an addiction, anxiety, and post-trauma specialist I know this is seen easily in people who struggle with substance use disorders, anxiety, and addictions to relationships, food, phones, video gaming, gambling, and screen time. We avoid feelings by doing, dodging, or denying them.

In my work, I find difficult feelings are suppressed or misunderstood because of four behaviours we experience usually at very young ages, and these affect our stages of emotional development: attachment, abandonment, acceptance, and approval. We usually experience too much abandonment, ineffective or nonexistent attachment, and not enough acceptance or approval from others and from ourselves. This multi-colored, mixed-up ball of behavioural yarn has us wound up internally, and likely we can't process or express difficult emotions.

If we first become aware that we have a huge smorgasbord of feelings to choose from, we can deepen our connection with how they affect us. If "we can name it, we can tame it" or seek to understand its origin and/or allow it to wave through us. However, if we keep ourselves too busy—if we cast these strong emotions off to the side or we are not even aware we have them—we return to the messages learned as a child or experiences in which our primary caregivers told us not to feel or we were feeling "too much." Often, we swat strong emotions away with mantras learned since childhood…messages such as, "you're too emotional," "suck it up," "get a grip," "think positive," "don't worry about it," or "it's no big deal." These statements and so many more give us one clear directive: don't feel our emotions.

I am a serial canner (my pantry holds nine months' worth of healthy foods every year). Stuffing, denying, dodging, deflecting, or using behaviours such as not setting boundaries, avoiding a conversation, being honest, assuming what another person thinks or feels without asking, or using alcohol, drugs, or destructive behaviours to avoid feelings reminds me of my pressure canner. Internal combustion and boiling temperatures have to be released and cooled in order for the canning process to be successful and the integrity of the contents to be preserved, just as they do with our emotions. If not, that pressure release transfers somewhere else—those feelings have to move to relieve the pressure! Our culture is in the "age of anxiety," but even deeper the cause, in my opinion, is the power, pleasure, and pain of avoiding feelings or actions which actually relieve emotional pain or suffering. There is pent-up pressure in avoidance of difficult feelings or situations of tension. As psychologist Dr. Susan David expresses in her TED talk, "Discomfort is the price of admission to a meaningful life."

Not doing, saying, or sharing our feelings can cause avoidance due to emotional discomfort for most of us. This is because of the fears we learn as a result of rejection, being judged or criticized, or worrying excessively about what other people think. As children, we learned effective defences to protect our feelings and avoid being hurt emotionally, mentally, or physically.

When our parents or care providers accused us of doing something wrong, we felt those "negative feelings" and therefore put up walls and beliefs that told us to never do that again!

Our culture does not clearly explain what stress or anxiety is, yet we experience these emotional clusters as infants, adolescents, and adults. In my work most people experience chronic stress or anxiety, substance misuse, or destructive behaviours because they have not developed emotionally. Their emotional growth is stuck in adolescence or earlier. I can almost always track avoidance and fears back to adolescence and childhood. It is a formidable time of development where we learn to belong, determine if we are enough, learn to people-please to become accepted, minimize our morals because we need approval, and attach to a peer group that grows us or holds us back but at least gives us a sense of belonging.

Stress can be naturally healthy in the case of situations such as being uncomfortable trying something new, reaching a goal, adding weights to your workout, or distances to your walking or running. Chronic stress stores in the body, potentially creating digestive disturbances, headaches, insomnia, or chronic muscle pains. Chronic stress and the effects of anxiety can be expressed by becoming easily agitated, frustrated, and moody. Maybe you feel overwhelmed, like you are losing control or need to take control. Clients share that they have difficulty relaxing and quieting their thoughts because they overthink and do not know how to shut or slow down all those racing thoughts. Chronic stress can manifest through worrying, inability to focus, changes in appetite such as either not eating or eating too much, procrastinating, avoiding responsibilities, or increased use of alcohol, drugs, cigarettes, or vaping. Nervous behaviours such as nail biting, skin picking, fidgeting, or pacing all indicate the body and brain are trying to find some sense of calm or release.

One client was addicted to sugar. She would go to the bulk food store, frantically buy thirty-five dollars of gummy bears, and sit, eating them just to feel something. After eight months of therapy and recovery from sugar and relationship addiction, she began to thaw out her feelings and identify them. She began to realize she was attracted to chaos and avoided calmness. The chemical surges had her engaged in lying, hiding, stashing, and going to different stores every second day so the staff at each store would not notice her frequency or amounts. Chaos had become more comfortable. After seven months of understanding her emotional causes of grief, loss, sadness, fear, guilt, and shame, she was able to lose 110 pounds, and with her doctors' help, eliminated all high blood pressure medication and antidepressants.

Experiencing stress or anxiousness every now and then is not something to be concerned about. Ongoing, chronic stress, however, can cause or exacerbate many serious health problems. Almost every client I have seen has various degrees of emotional and physical pain. Our brains and bodies are not separate. As in nature, all organs and ecologies are integrated and systematically connected.

Neuroscience and developments in the field of psychological and physical pain are exciting because they offer ways we can begin to understand, manage, and possibly become pain-free. Dr. Rachel Zoffness is a pain psychologist and an Assistant Clinical Professor at the UCSF School of Medicine, where she teaches pain education for medical residents. She is author of *The Pain Management Workbook*, an integrative, evidence-based treatment protocol for adults living with chronic pain; and *The Chronic Pain and Illness Workbook for Teens*, the first pain workbook for youth. She also writes the *Psychology Today* column "Pain, Explained."

Dr. Zoffness shares the science of physical pain. She says, "What science tells us is pain is not purely biomedical. It's actually this different and more complex thing, which is biopsychosocial." Following are additional observations from Dr. Zoffness:

- Pain is complex, and doing one single thing over years and years that has not worked, is probably not the right way to go.
- Pain is never purely physical. It's always also emotional.
- Unless we're taking care of our thoughts and emotions, we're actually not really treating this thing we call pain effectively.
- Ninety-six percent of medical schools in the US and Canada have zero dedicated compulsory pain education.
- Pain, by definition, is a subjective experience.

On treating chronic pain, Dr. Zoffness says, "Keep doing exactly what you're doing and follow your gut. Trust your intuition, and know that following the path of the thing that you love is the thing that's going to bring you to where you need to be professionally."

Consider a client of mine, Janine. She had initially come to my office to attend our intensive outpatient program for alcoholism, but she also has chronic back pain, fibromyalgia and migraines. Janine was a mother of three adult children, had been married twenty-five years, and operated a successful business. She had not drank alcohol for twenty years but had recently gone back to drinking casually, which turned progressively into a daily bottle or two of wine. In our initial visit, she shared that for all her life she could not "get out of her head" and she could not stop, "thinking." She had been on high doses of antidepressants for thirty years. She spoke very quickly and anxiously, she exercised obsessively, never stopped in her day, and only slept four hours a night. Physically she could identify with the exhaustion she felt but instead of resting, she over-exercised, obsessed about how she looked, and overworked. She either had chronic constipation or days of diarrhea.

We determined her defence mechanisms, which included minimizing her feelings and intellectualizing her explanations instead of connecting feelings to her thoughts or body. She began to create a new vocabulary of identifying her emotions and expressing them. Instead of simply saying she was "stressed or anxious," I shared with her a list of emotions she could begin to connect with and identify. "I'm anxious" became, "I'm worried," "I'm afraid," "I'm uncertain," or "I'm angry."

As her body and emotions began to share and express her pain, Janine was able to calm her thoughts and her body.

By beginning to connect thoughts and feelings, Janine could bravely begin to open up and share her childhood traumas of physical abuse. Most people minimize feelings—it is a way to prevent more discomfort. Over many months of one-on-one sessions, Janine cried and released anger, shame, and guilt. Sadness poured out, and she grieved. Her emotionally lost inner child began to feel. She shared the disbelief of how deeply she had unconsciously minimized painful events of her past. Many months later she had no symptoms of fibromyalgia. She had not experienced a migraine headache for months, and she was walking two hours per day with no physical pain.

I will never ask a client to change their story. I invite them to connect to the forgotten, hidden, or buried feelings which help people thaw and release their emotional pain. Usually, a side effect is some form of physical release. It is here where Janine found so many incredible qualities about that lost kid and how she was able to find a way through traumatic circumstances, resolving her chronic stress and anxiety responses fifty years later.

There is a powerful transformation in listening or returning to painful stories never felt. We are required to stay in the expression of the experience, but real emotional release can come from returning and caring for the emotional entrapment within the stories or histories that were never told. We reconnect to the child and progressively build the bridge to adulthood. Janine literally returned to the young girl she left behind. Emotional pain from a survival perspective has a way of being avoided because we have usually experienced emotional pain without any comfort, soothing, or care.

As in traumatic responses at various degrees, we are left alone having not been supported emotionally, mentally, or physically. As a result, the trauma is stuck or locked in our bodies. It can be stuck in the vault of our entire nervous system, and we carry it with us our entire lives. We carry our stories or experiences in every cell of our body. The psychological, social, and physical aspects of every experience are the triads of our life stories. Untangling the emotional knots gives our histories self-compassion and deep understanding. We never escape them; we may try to temporarily relieve them or avoid them, but the deep-rooted therapeutic approach seeks first to understand and soothe them…just as they could have been but were avoided. Simple awareness as to what was occurring never happened.

There is a consistent feeling clients express throughout this process: relief. Clients experience this feeling in their entire body. Sharing pain and connecting to all the feelings associated with that pain is liberating. Janine has reduced her antidepressant medication substantially, and her anxiety symptoms are gone after several months of working with me.

If there are any doubts that emotions do not exist in our physiology or that specific emotions do not exist, you may find this interesting. Cited from an article in *Smithsonian Magazine*, written by Joseph Strombery, he writes about photographer Rose-Lynn Fisher capturing the tears of grief, joy, laughter, and irritation in extreme detail. Rose-Lynn Fisher created a multi-year photography project in which she collected, photographed, and examined more than a hundred years from herself and a handful of volunteers, including a newborn baby.

All tears contain a variety of biological substances (including oils, antibodies, and enzymes) suspended in salt water, but as Fisher saw, tears from each of the different categories include distinct molecules as well. Emotional tears, for instance, have been found to contain protein-based hormones including the neurotransmitter leucine enkephalin, a natural painkiller released when the body is under stress.

Additionally, because the structures seen under the microscope are largely crystallized salt, the circumstances under which the tear dries can lead to radically dissimilar shapes and formations. So, two tears with the exact same chemical makeup can look very different up close. "There are so many variables—there's the chemistry, the viscosity, the setting, the evaporation rate and the settings of the microscope," Fisher says. Closely studying tears for so long has made Fisher think of them as more than just a salty liquid we discharge during difficult moments. "Tears are the medium of our most primal language in moments as unrelenting as death, as basic as hunger and as complex as a rite of passage," she says. "It's as though each one of our tears carries a microcosm of the collective human experience, like a drop of the ocean."

Dr. Michael Roizen of the non-profit medical center, Cleveland Clinic, said that crying for emotional reasons, not only releases tensions, but it can physically deliver nutrients and "wash out stress-related toxins." Crying also helps us communicate our needs, like distress or sincerity. Each type of tear—basal, reflexive, or emotional—has its own specially designed role, each of which the Cleveland Clinic says is produced by glands around the eye to maintain eye health.

Dr. Roizen explains emotional tears. "These tears gush in response to strong emotions like sadness, grief, joy, or anger. They all contain the same chemical makeup, but more stress hormones and natural painkillers than other kinds of tears. Humans and animals have compounds in our body fluids that give off subtle messages to other members of the species. That's why sometimes our tears can relay chemical messages (either intended or unintended) to someone close by like, "stay away," for example."

According to the American Academy of Ophthalmology, emotional tears are triggered by empathy, compassionate and societal pain, physical pain, attachment-related pain, and sentimental or moral feelings. Researchers at Harvard medical school have established that crying releases oxytocin and endogenous opioids, also known as endorphins. These feel-good chemicals help ease both physical and emotional pain.[4]

Popular culture, for its part, has always known the value of a good cry as a way to feel better—and maybe even to experience physical pleasure.

I believe emotions and science will continue to grow and integrate in their understanding and clinical research. Everyone has feelings and the relationship to longevity, health, wellness or the effects on every aspect of ourselves is under-researched.

Psychoneuroimmunology has been a respected field of study for almost forty years, but we simply do not find research in our day-to-day lives without really looking for it. Intuitively I believe we know emotions are connected to every cell of our bodies; our emotions are with us constantly in everyday life. Feelings are neither right nor wrong, they just are. Many of us have been taught to categorize our feelings as good or bad, positive or negative. This categorization can be destructive because it reinforces the idea that some feelings aren't acceptable and perpetuates the directive of "don't feel." It's more helpful

---

[4] https://www.aao.org/eye-health/tips-prevention/all-about-emotional-tears

and often a great start in learning about our feelings to rather think of feelings as comfortable or uncomfortable because most people can rarely name more than four or five emotions. Using these terms rather than "good" or "bad" serves as an easier starting point for people who have trouble identifying their emotions. Typically, uncomfortable or negative emotions are classified as sadness, anger, guilt, shame, remorse, grief, or fear. We all want to feel pleasurable feelings such as joy and confidence, but feeling angry, sad, or fearful are numerous emotions many of us avoid.

Our feelings come from two different places:

- External Reality—what is going on around us and to us.
- Internal Thoughts—our perception of the way life is going or how it affects us.

Feelings can be more powerful than our thinking. By avoiding feelings, we do not realize they need to be dealt with, but emotional outbursts, bodily tension, anxiety, depression, or isolation can all be signs that this is exactly what is happening. When we begin to learn this, it can feel difficult when feelings start to come out of "hibernation." They are asking to be noticed or acknowledged.

Most of us have not learned or been taught how to grow into emotional maturity and fluency. Our emotional development is arrested when we begin to protect ourselves emotionally from others, or we put too much focus on caring for the feelings or needs of others— that we lose the recognition of our own. We often find it difficult to cope with feelings of anger, fear, loneliness, guilt, sadness, and more. Recovering from trauma, anxiety, addictions (relationships, behavioural, and substance misuse), and overdoing begins, and we continue to learn to listen to feelings and figure out how to cope with them.

The way we deal with our feelings is the most frequent source of difficulty in our relationships with others. Although each of us continually experiences feelings about others and ourselves, most of us have not yet learned to accept and use our emotions constructively. We not only are uncomfortable when others share strong feelings, but most of us do not even recognize, much less accept, our own feelings. We know intellectually that it is natural to have feelings. Yet we believe feelings are disruptive—the sources of obstacles and problems in living and working with others—instead of the key to calmness and emotional fluency or regulation.

It is not our feelings that are the source of difficulty in our interactions with others, but the way we deal with them, such as our avoidance to expressing them in a healthy way. Because of our fear and discomfort with feelings, we spend much effort and vital energy trying to deny or ignore them. Look around you and observe how you and others deal with feelings.

What do we say to our feelings and to other people's feelings? We judge ourselves and our feelings harshly. We say things like:

- Don't be so emotional.
- Don't be angry.
- Don't be so negative. Think positive.
- Don't be such a victim.
- Get over it. Move on.

When we speak life into these statements and direct them toward ourselves and others, what we are really saying is "don't feel." Then we disconnect from feeling. I say to clients that "if we can't name it, we can't tame it." Instead of expressing our feelings in a healthy way, we people-please and discount or deny our feelings. We are honest people, yet dishonest as to how we really feel. The need for approval and acceptance temporarily puts our real feelings on the shelf as the fear of what others think takes over.

Emotions move. In general, the closer the feeling is to the present moment, the more real it is—the more difficult it is to open and discuss. If we are not used to feeling the discomfort those emotions give us, we tend to defend them by minimizing them or not even being able to name them or connect to that emotion.

In my office, I have a feelings wheel that really helps people name and identify their feelings. The wheel has 130 emotions a client can connect to, begin to relate to, or consider feeling. The important question is, does it matter whether we discuss our feelings as we are having them?

The answer is yes. Feelings we try to bury inside do not disappear. They tend to stay stored up until, sooner or later, they find a way to come out—either by internally abandoning ourselves or externally impacting others through our impulsive and reactionary words or the common silent treatment. The problem is that when feelings have been simmering for a while, they tend to come out in inappropriately strong ways or focus on the wrong target, inwardly or outwardly. Most of us have said things we did not mean that were hurtful, but we also do not say how we feel in an honest and respectful way in order to create more intimacy in our relationship with self or others.

Rather than deciding not to talk about your feelings, decide when would be an appropriate time to do so, and with whom? Talking about your feelings is not only right, but also extremely important. We are emotionally fluent when our feelings, words, and behaviours all match. Examples include:

- I feel sad because my dog died.
- I feel like crying, and I could really use a hug.
- I am so excited today, because I am buying my first home, but I am also feeling nervous because it's new and I am out of my comfort zone.
- I feel lonely, so I am calling a friend to see if he/she will come over to watch a movie.
- I am angry and sad that I cannot drink alcohol anymore.

Learning emotional fluency is important for recovering from anything! What is it we could share to prevent or release the potential for trauma, addiction, or unhealthy relationships? Emotional fluency builds hope and a solid and secure road to experiencing a healthy relationship with ourselves and others, but it requires us to have our feelings on hand and begins by identifying a range of options. The blessings of this process are to give the body and brain a place to begin again; a place to process where the emotional suppressions started. Then the emotional energy is to be moved into a more present experience both in the brain and the body. For example, I do not need to stay angry and self-destructive about what harmful things or experiences were done to me, but my emotions do need a place to be heard and therefore soothed and brought into current life experiences.

As a child, I was never able to talk about anger, and I certainly learned it was "bad" because anger was, in my world, historically expressed through physical aggression and verbal, emotional, or mental abuse.

In my family of origin, we could not say we felt hurt or sad because the consequence of that perceived weakness was to get punched, disapproved, or subject to ridicule. It was a good idea learned by trying to survive and through self-protection to not express those feelings! Later in my adulthood when my wasbund (my name for my two former spouses) sought a sexual relationship outside our marriage, asked for a divorce, and then declared bankruptcy, it was too much for my childhood emotional experiences. I involuntarily stuffed my feelings about it all deep down within me because anger and feelings of abandonment and betrayal meant more pain. This is what we did when we were hurt in my family of origin.

My emotional immaturity regressed and turned into depression and increased alcohol use. I discovered emotions, and I reiterated to myself those learned messages: don't feel your emotions, don't trust them, and definitely do not talk about them…that means more pain! All of this creates unhealthy relationships and bypasses the basics of honesty and intimacy. Marriages and relationships that fail or are stagnant or unhappy are because the language of emotional identification or healthy expressions and acceptance of these feelings are as foreign as another planet and usually scary or unacceptable on multiple levels. Instead of getting into discomfort or knowing how to utilize the courage and vulnerability which leads to resolution, it's easier for people to leave or avoid conflict in the relationship.

My childhood taught me that the chances of ever finding an intimate partner who wanted to understand my feelings, understand their feelings, or stay long enough to work through them, was going to be highly unlikely. That proved to be true.

Similarly, to my family of origin, the men I married simply left for a more "positive" relationship or the ease of getting a new relationship was the priority, not emotional intimacy or conflict resolution. Quit. People give up too soon, unwilling to change—mostly because they cannot stay in the fire of discomfort and rigorous honesty. But the ones who do take their relationship into a whole new level of fulfillment, joy, transparency, honesty, and intimacy can and do sustain a lasting love of acceptance and discovering the differences can also be the strengths. I have a decent relationship with failure, I accept it is necessary. Those marriages taught me to have a relationship with myself first—the one connection I was not taught to have as a child or young adult. Those relationships helped me to understand that by growing myself up emotionally, I could discern what a healthy relationship is and find others who could do the same.

When experiencing the emotional aftermath of divorce, I learned through a great therapist that my stronger and more intense emotions had a voice. She said to me, "Dawn, you have the right to be angry." I do!?" That was the first time I had ever heard a moment of validation. It didn't mean that I had to act out the anger. What was more powerful was someone actually listened to how I was feeling without telling me it was wrong or fixing it. She continued to say, "Perhaps aggressive and abandonment behaviours in your family taught you not to have any right to feel certain feelings. Maybe emotionally unavailable or passive partners were on some deep level safer for you?"

I became aware that in my family of origin sharing emotions were meant with contempt, criticism, withdrawal, leaving the room or conversation, defensiveness, and eventually avoiding them until people left the relationship entirely. Not being able to healthily express anger or frustration (or any suppressed or misunderstood emotions) continues to give reinforcement to our primal fears such that I am not good enough and I am not loved.

Emotions are powerful because they are rooted in the meaning of words, beliefs and experiences. Our emotions create beliefs and behaviours, and our beliefs and behaviours reinforce those emotions. It is all a misunderstanding or a constructive or destructive emotional cycling until we seek to understand a renewed meaning.

Once I learned that anger tells us boundaries have been crossed or violated, our needs are not being met, and we perceive a threat in a relationship with others or in our body, I began to understand how anger is a healthy protection strategy! Once I was able to recognize, accept, and share the honest feelings associated with my anger—feelings such as sadness, grief, frustration, and fear, I was able to turn the energy of my anger into action.

I learned the power of that anger would be converted into my own growth and goals. I moved all those emotions by talking about them, moving my body with them, understanding they were a healthy expression, and using the strategies and principles in this book. My body and brain began to settle. Peace found me through the expression of honest and real emotions. It took me two years. That was decent, considering I had decades of training. With practice and radical honesty, my emotional, physiological, and psychological bank account learned the power of compound interest, and I became emotionally rich and regulated. But being raised to not understand my emotions and being with people who did the same had its costs. For me, the greatest cost was the effects divorce had on my children... the same pain I experienced in my childhood. The trauma and behaviour I learned from my family of origin had seeped and transferred their way into another generational bloodstream.

My father was married seven times, my mother three times, and none of those marriages were happy ones. I never set out to destabilize my own children. In fact, I dreamed of creating the family life and safe and secure environments I never had. My two failed marriages had a cause and effect on my own children, as it has with every child, and that potential for pain or trauma should never be minimized. It never leaves you, nor should it. It has been the pain from my childhood and the people who surrounded me which fuelled my own recovery and inspired me to make it right for myself and my own children. Through challenges, I found ways to heal trauma with the hopes I could shine some light for others to do the same.

The struggle is a worthwhile building block, but suffering is not necessary with knowledge, awareness, and willingness. The loss of our family foundations and belonging to a sense of worth and safety is deeply rooted in our stages of development, or lack thereof. It is my hope that we can find the bravery and willingness to change generational patterns and needless suffering by simply becoming emotionally healed people. I believe with awareness, guidance, wisdom, and willingness each generation can truly improve and recover.

We have many feelings, and each has various hues and shades. We generally need to develop healthy identification and expressive relationships with such feelings as fear, anger, guilt, shame, and joy. Because our survival brains look to avoid pain/discomfort and seek pleasure, it is the difficult, stronger, or so-called "negative" feelings that we have learned to avoid. Yet, each of these feelings, when embraced and expressed, offer gifts…gifts we would never receive if we did not dig deep into our roots and learn how to understand, accept, or change them.

FEAR: The Gift of Fear is Wisdom and Freedom.

Embracing our fears gives us the opportunity to notice what is life-destructive and what is life-embracing. Both involve risk. Fear gives us pause to reflect and make evaluations.

We gain wisdom from listening to our fear. Freedom is the ability to choose to use that wisdom.

Frank, a client of mine, was comfortable having many female friends. Based on his father abandoning him at fourteen years old, his fear held him hostage from creating trust with other men, and therefore trust and freedom within himself. Drinking with them was no problem, but feeling accepted by another man was inconceivable to Frank. He was very resistant to change (fear), but with his willingness to gently overcome his fear, he courageously began attending a men's emotional support group. The gift of fear shared with him that he could have meaningful relationships with men and that other men shared what he felt.

There is tremendous stigma about men having feelings (or women being too emotional). The gift of fear was his awareness of what he was afraid of—vulnerability, being judged, and being deemed not good enough. The freedom to now have meaningful relationships has far surpassed his greatest fear. He could now grow into that fourteen-year-old boy, connect to the abandonment he suffered, and rise above it to retrieve what had been lost.

ANGER: The Gift of Anger Is Strength and Protectiveness.

Our anger is a source of energy and strength which enables us to create change. It helps us to defend and guard what we value. Anger can become displaced and also be a protector from sadness and grief. We may have learned and believed that showing sadness or crying was wrong or we were judged harshly at some point for showing those feelings. Many of us have not been able to feel sadness, share, or shed a tear. Perhaps we have learned it is a sign of weakness to emotionally expose ourselves. Sadness and grief can smolder in ashes of loss, and that pain then shape-shifts itself into an energy that has to be released or expressed. This can be communicated constructively or destructively, of course. Anger can exercise our voices of abandonment or injustice as it is a highly charged voice that deserves to be heard, understood, and respectfully shared with others. Anger can create hurt or harm, and releasing anger in a healthy way can also give relief and build assertiveness.

Kim was a woman who was worried about her anger. She had physical outbursts of pushing and shoving toward her husband during her marriage. Now that she was divorced, she felt afraid this anger could come out again. Kim shared a history of becoming sexually active at thirteen years old. She was living in a physically violent home, and her relief and temporary sense of belonging were found sexually. She identified emotionally that this was the attention and approval she needed while living in a volatile family home.

Gently, I asked Kim to use her imagination to return to the home she lived in as a child and asked her to describe what it felt like to live in that environment. Like most of us, Kim initially intellectualized what she was seeing, and it often takes a few minutes to feel safe enough to trust our feelings. It can often take a few minutes or using the feelings wheel to identify or trust our feelings. Kim had never returned nor ever considered visiting that little kid who had been wounded in the past; she had put her out of her mind and certainly out of any connection to feeling or understanding what she went through. Waves of tears began to come as I gently guided Kim through the process of returning and observing what life was like for her little kid. Kim began to hear and to connect to the voice and feelings of her little kid who never had anyone to share with…By sharing with someone safe, Kim began to understand so many of her life patterns. By returning to the wounded child, she could offer her compassion, understanding, and care, exactly what we all need at the most impactful times in our lives.

After the exercise was over, she shared a statement and an emotion that 100 percent of my clients express. "I feel relieved." Kim said, "I had no idea I had forgotten her and what she had been through." Returning Kim to her younger self allowed her to grow back into her fifty-five-year-old self over the process of three months. Slowly and patiently, her younger self could finally share all the qualities she had become as a result of her traumatic past. Kim softened as equally as she released anger and sorrow. Identifying her anger at her mother who abandoned her, her sister who hit her, and the shame of believing she was a failure in school, Kim was able to soften into the sadness and the loss of her little thirteen-year-old self. Over numerous sessions, her anger opened the door to reconnect her to the lost child she had forgotten but could now care for. She has not left her since and now cares for her as often as she needs to.

Anger is one of the most misunderstood emotions. More times than not, people defend themselves and preserve their sense of identity by saying "I am not an angry person," "We never argue," or "I am always a nice person." To them I say, "With your willingness and curiosity, if you stay with me long enough, we'll find your anger, wisdom and freedom.

SHAME: The Gift of Shame Is Guardianship and Honesty.

Since our feelings are healthy, shame, too, must be healthy. Based on developmental milestones, shame starts to show up in children as young as one to two years of age.

Do you remember being three years old? Most of us have a few vague memories of this time period. You may remember exploring your world and wanting to do things on your own. Perhaps you're a parent yourself and have noticed your toddler beginning to express more independence and develop his or her own interests. This willful exploration and sense of independence is called autonomy, and it's a notable feature of the toddler years.

Parenting plays a large role in a child's development, and how a parent handles this tumultuous and exciting stage can have a lifelong impact. Developmental psychologist, Erik

Erikson (1902-1994), referred to the conflicts of this life period as autonomy vs. shame and doubt, which is one of eight stages in his theory of psychosocial development. In order to fully understand this second stage which examines autonomy vs. shame and doubt, it would be helpful to view it within the context of Erikson's theory as a whole.

Erikson's Theory

You've likely heard of Austrian neurologist Sigmund Freud. His writings on child development focused on the influence of early experiences on personality. He conceptualized development as stages, occurring during the first five years of life. Each stage is characterized by conflict that, if left unresolved, could lead to psychological dysfunction later in life.

Like many of Freud's contemporaries and future scholars, Erikson was influenced by Freud's work. He also viewed development in stages that coincided with different types of conflict, but he placed a strong emphasis on social interaction. Erikson's stages span a person's entire life rather than the first few years. Erikson's theory consists of eight stages, written as conflicts that people encounter at certain age ranges. The following are brief descriptions of each stage to give some context:

- Trust vs. Mistrust (birth to one year): Infants rely on their parents to provide care, and they develop trust when their needs are met. Otherwise, they may grow up to be mistrustful and suspicious.
- Autonomy vs. Shame and Doubt (one to three years): Toddlers express willful exploration. If they are restricted

by controlling parents, they may experience shame and doubt.

- Initiative vs. Guilt (three to five years): At this point, children develop an idea of who they are through play, interaction, and self-expression. Too much criticism can lead to guilt.
- Industry vs. Inferiority (five to twelve years): In this stage, children learn to read and create. Those who are positively reinforced will feel industrious, whereas those who are stifled may feel inferior.
- Identity vs. Role Confusion (twelve to eighteen years): Teens seek to understand their identities and the roles they will play in society. Failure to do so can lead to confusion.
- Intimacy vs. Isolation (eighteen to forty years): This stage is marked by seeking intimacy with others. If this is not successful, a sense of isolation can develop.
- Generativity vs. Stagnation (forty to sixty-five years): In middle adulthood, people focus on establishing careers and future generations. Failure to do so can lead to a sense of stagnation.
- Integrity vs. Despair (sixty-five plus years): Older adults often contemplate their lives and accomplishments. If they feel successful, they have a sense of integrity. If not, they feel despair.

Shame arises when we feel our privacy is violated and/or we feel exposed, especially when others find out we have done something wrong. However, the unique experience here is that many times we did not do something wrong, but it was assumed we did. Another way to potentially feel shame is feeling or thinking something that people we admire believe is wrong or immoral.

Shame could also manifest by comparing some aspect of ourselves or our behaviours to a standard and feeling like we do not live up to it. When we're betrayed by someone we love or laughed at and in front of others, we also feel shame. Embracing and sharing our shame is our most valuable path to intimacy. There is no intimacy without vulnerability. We are never more vulnerable than when we tell what happened to us that made us feel bad about ourselves. Sharing our shame allows us to be loved and love more deeply. A healthy sense of shame gives humility, honour, and pride. Our feelings of shame may originate from something we actually did that was wrong or something a person accuses us of or blames us for that we had nothing to do with. A lot of our shame is actually not ours, but we were led to believe it was.

All of these emotions feel uncomfortable, but protecting them or denying them keeps us disconnected from ourselves and others. Vulnerability risks judgment from inside and out. If we were a beautiful puzzle, fear of judgment from vulnerability causes us to miss the pieces needed to create completeness or wholeness.

One main factor key to renewing or restoring a healthy relationship with our emotional selves is emotional honesty, which takes practice to learn. Emotional honesty certainly does not come naturally if we have avoided or not trusted others or ourselves with our feelings. Instead, it is learned and earned through practice in identifying what emotions we feel and then being able to consistently trust them as being accurate or congruent. It takes practice because we were taught mostly by our caregivers or people we gave authority or trust to about what we should or should not feel. Parents, caregivers, coaches, teachers, colleagues, friends... all of these relationships shape our perceptions. This can often lead to us defending what we have been taught and believed as "good or bad" feelings. If they're "bad," we defend them as a way to avoid rejection or criticism. Emotional honesty or identification of feelings in most of us need a language and a connection to hook on to.

Patrick was a sixteen-year-old man who came to see me for high levels of anxiety. His parents and teachers were concerned about him spending a lot of time alone, and he was missing more school because he was not feeling well. Yet, there were no pathological reasons anyone could find. Patrick lived on a family farm, and his story led me to examine when he accidently left a gate open and his brother's horse got out. As a result, the horse seriously injured itself, falling down an embankment. The horse had broken its back and had to be put down immediately. Patrick felt shame for his mistake, so he internalized shame into unexpressed anger at himself and a deep sadness for what he had done wrong. He loved his brother, and he loved the horse. His despair and shame were converted into self-loathing, and his emotional unresolve deepened over months into isolation and depression.

Patrick shared that his family never talked about the incident. "We do not talk about feelings," he said. But once Patrick could release and share the story and his feelings over time in our sessions, he was the one who courageously and vulnerably shared those same feelings with his brother. Patrick apologized for accidentally leaving the gate open. Both he and his brother openly cried together and through the gift of shame could begin to feel the grief and sadness they had both felt but buried for more than five years.

It's not always easy to reduce feelings of shame, especially when shame isn't fully understood. People sometimes feel ashamed of some part of themselves without knowing why. A person might also feel shame when other people know about the actions that they feel guilty over.

Shame can cause people to feel unworthy or somehow inadequate. It may lead to isolation, acts of self-punishment, or other potentially harmful behaviours.

As with all emotions, shame can be a destructive cyclical process, and at some point, it really asks to heal once and for all. One of the most destructive fears is that people are not "good enough," and it is rooted in shame. Shame can be misdirected and turned inwardly with needless suffering as the consequence. Our undesired or perhaps pathologized behaviours often stem from emotional and psychological misunderstandings.

GUILT: The Gift of Guilt Is the Ability to See We Have an Impact on Others.

There is a part of guilt that is healthy; a sense of wrong and right, a sense of knowing that some things are okay, and others are not. Guilt says we did something wrong, we have hurt someone or ourselves, we are accountable, and we apologize. Guilt could attend to the parts of us that know we have done something "wrong" or something that is not in alignment with our values or the type of person we want to be. Guilt can be the stop sign to shame. Most people cycle between guilt and shame, but unnecessarily. We don't need to feel the debilitating effects of shame if we can keep our guilt in check.

Guilt can be about what we do, while shame is often described as being about who we are. Guilt is the ability to see the consequences of our behaviour.

Guilt and shame are often confused with each other, though they're distinct emotions. Guilt can feel regretful or remorseful. People may feel guilt over things they actually did wrong, things they believe were their fault, or things they had no responsibility for. Survivor's guilt, for example, can affect people who survived tragedies when many others died or were harmed.

People tend to only feel guilty over actions they see as wrong. A person who believes they are entitled to a higher wage may steal small amounts of money from their boss without ever feeling guilty. But a person who finds a wallet and keeps the money inside without making any effort to find the owner may feel guilty for months or even years, especially if they believe the "right" thing to do would have been to turn in the wallet.

Some people experience chronic guilt, which can lead to feelings of repetitive shame. This type of guilt can lead to destructive actions instead of positive change. People might also manipulate others with what's known as a "guilt trip" by using a person's guilty feelings as a tool for them to do what they want.

Someone who feels guilty about something they did might take steps to correct their mistake, apologize, or otherwise make amends. This usually causes feelings of guilt to subside.

But shame can be burdensome as a sense of responsibility that relates to the self, so it can be more difficult to address.

Guilt is a conditioned emotion. In other words, people are conditioned (they learn) to feel guilty. As we see in Erickson's stages of development, certain factors may make it more likely a person experiences chronic or excessive guilt. These factors might include their culture, family, or religious upbringing. If parents consistently make a child feel guilty or consistently withhold praise, for example, the child may come to feel that nothing they do is ever good enough. This can lead to a guilt cycle.

People who struggle to overcome feelings of chronic guilt may have a higher risk for depression, anxiety, or other mental health concerns.

When we do not learn to feel or understand our feelings accurately, we assume way too much responsibility for other people. If we can admit we did something wrong, guilt can ease its grip. If we are over-responsible and believe we are the cause of other people behaving in harmful or destructive ways, we suffer emotionally and psychologically. We hear this consistently from people who have experienced psychological and abusive traumas. This type of guilt is a very close cousin to shame. It is learned mostly through our assumptions or by people telling us and, in turn, our believing that we are to blame for their feelings. This projection from others is inaccurate. This deeply misunderstood guilt can be resolved with a great therapist by returning to events or timelines of our past, as we did with Kim. Many of our suppressed feelings and misinterpreted identities are not our fault—we learned them through others.

Discerning how we feel is the first step. It's important to determine if the guilt is our responsibility. and if we need to apologize or change. Or, are we assuming way too much responsibility for another person? The feeling of guilt that helps us be aware of our actions and how they affect others can encourage us to improve or adjust our behaviours.

If you have never been able to sense or feel guilt, chances are you have denied, dodged, or distracted it. Not being aware or accountable for guilt can show up in anxiety, depression, addiction, and self-harm. Remember, guilt and all its behaviours can be learned constructively or destructively. Of course, it can be easier and more worthwhile to attend to our guilt while being aware of these behaviours or thoughts. For example, you may feel guilt when doing or thinking something you believe is wrong, doing or thinking something that violates your personal values, or not doing something you said you would do.

One common behaviour that we all do to not feel guilt is avoidance. Once we become more assertive and emotionally honest by recognizing, accepting, and sharing our feelings, guilt will dissolve, and it won't hold our self-esteem or fear hostage any longer.

SADNESS: the Gift of Sadness Is Healing and Empathy.

Sadness is a healing feeling. Our ability to experience sadness allows us to heal from loss and hurt. Sadness is the key stage in the grief process that precedes acceptance and serenity. Most of the people I see have not become aware that sadness and feeling our grief is natural and relieving. I believe one of the most beautiful parts of sadness is that we remember. We may remember the parts of our lives or people we have lost. When we begin to feel safe and accepted, we are able to stop minimizing or feeling shameful for those feelings.

We may share sadness about losing our childhood, our parents' divorce, moving away from friends, or the loss of a pet or someone we cared for. When we can talk about how we feel without the experience of critical judgment, the embarrassment, shame, or guilt usually alleviates itself, and a wave of relief comes over us.

A client who was unhappy in her marriage shared that after she could talk with someone without being told what to do or how not to feel, she actually felt as if a thousand-pound weight had been lifted off of her. We cannot heal what we do not feel. As a therapist, the power of listening to another person's loneliness, sadness, fear, or shame is often the only medicine a person needs to feel better and move forward. Unpacking the heaviness we carry builds strong roots in our acceptance for all our feelings.

Even years later when I see a golden retriever, I think of my golden retriever, Milo Man. For fourteen years Milo was my constant companion. Some days any dog reminds me of my life with Milo always by my side, and sadness can come over me without warning. I let it. Sadness or grief may never go away, for love and missing companionship and connection are never lost in forgetfulness. The sadness over a lost childhood or the family or relationships we lost can convert and integrate our own empathy and the healing journey to feel emotions. We can accept ourselves and others by no longer denying our feelings but sharing them so we can create more beauty, belonging, acceptance, and love.

The gifts and awareness of our feelings have been seeking your attention, constructively or destructively. Emotions can be wise and wonderful teachers.

Write about the experiences of when you did listen to these emotions and when you did not? When did you first learn the emotion? When did you not listen to it and why? When did you utilize this emotion to your advantage? When did sharing this emotion create a greater connection to yourself or another?

Fear:

_____
_____
_____
_____

Anger:

_____
_____
_____
_____

Shame:

_____
_____
_____
_____

Guilt:

_____

_____

_____

Sadness:

_____

_____

_____

Behaviourally, feelings can be a constructive or destructive pendulum. We can stay in the constrictions of sadness, fear, anger, or despair, avoid what those emotions may be sharing with us, and likely head into an unacknowledged state of depression. Or, we can avoid those emotions entirely and live our best life in a state of denial, swinging the other way on the pendulum and only choosing to feel bliss, pleasure, and feast on dopamine, oblivious to any consequences of staying on the emotional high side.

My area of expertise is in behavioural addictions, relationship dependencies, and post-trauma. Addictions and traumas of all kinds are formed because of this pendulum swinging in the repetitive direction of the emotional highs and lows. Without exploring and being curious about the variety of emotional waves, we never find the ability to weather those waves or find calm in the storms. With practice, we can find our rhythm or homeostasis by acknowledging and recognizing our feelings coming in and out of the shoreline. The emotional tides will ebb and flow. Our only responsibility is to do our best to identify with as many feelings as we can, talk about them with someone we trust, and accept our feelings as neither good nor bad. Moving those feelings that stay stuck on either end of the pendulum will be revealed in the next chapters.

Truly, there is no way we can thrive in the bliss of affection, empathy, tenderness, excitement, peace, joy, or love without being open to connecting and experiencing anger, fear, jealousy, guilt, embarrassment, frustration, grief, and even hatred. If we want a life in which we thrive rather than wither needlessly, we must be willing to accept the abyss or disastrous results as well as the victories. As I see it and as my clients express to me, the deeper darkness or abyss is avoiding feelings and then expressing them to a point where we take no risks, play it safe, and deny there are any problems.

In avoiding feeling and change, we eventually lose the most important nourishment—relationship with ourselves and others. A life of reverence is rich with meaning, emotional fluency, and one that not only fills your own coffee cup to the brim but also invites a waterfall of emotional connection which spills over, filling the cups of others who were moved or inspired by your emotional honesty. When we begin to practice emotional honesty, we become guides for deep connection and courageous vulnerability. It's essentially an emotional adventure!

The Emotional Medicine Wheel

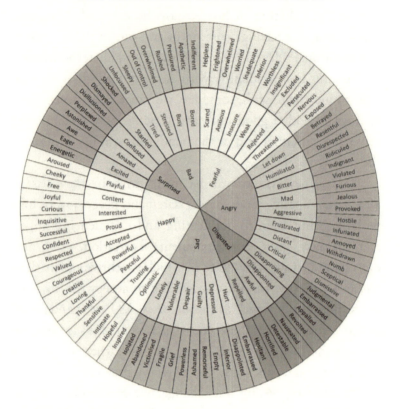

With practice, eventually we will be able to winter our darker or heavier painful emotions and simply accept them. Once we realize that, as in nature, emotions are seasonal and cyclical. We can weather the storm of changing feelings. We begin by creating a healthy relationship with our emotions, which, in return, loosen their grip on our thoughts of catastrophizing, futurizing, making the past the present, and insecure or terrifying assumptions of all kinds.

Our emotions can be our gifts or curses. With the pendulum swinging too far one way or the other, feelings can keep us clinging in the rotting roots or grow us into solid and strong cedar trees. Be free to feel and allow yourself to nestle, to move, to shout, to sing, to share, to shine, but remain rooted. Not recognizing, accepting, and sharing our feelings with people we can't trust or feel safe with is the deepest cause of unresolved pain. People often try to "fix" or judge our feelings.

Emotions are the shackled or silenced inner voices waking you toward recovering from trauma or rooting you to regulate and heal from traumatic responses. Actively listen to your feelings—all of them. They are in every cell of your body and in every breath of the past. They are your history but often require an updated edition. Now is the time to move them. Sing, dance, sit, share, nestle, move, holler, howl, connect, create, cry, write, build, and give your emotions exactly what they ask of you. Be free to feel and allow yourself to shine, but remain rooted. Emotions are you and your voice graciously or bravely calling you to be loved and live, authentically.

> *"As long as the complex remains outside of awareness, we will find ourselves acting out of compulsion, reacting to scenes in our life with the same consciousness that was traumatized in the first place. What we seek is the ability to encounter life openly, freely and with soul. We cannot control what comes to us, what moods arise, what circumstances befall us. What we can do is work to maintain our adult presence, keeping it anchored and firmly rooted. This enables us to meet our life with compassion and to receive our suffering without judgments. This is a core piece in our apprenticeship with sorrow."*
>
> —Francis Weller, *The Wild Edge of Sorrow: Rituals of Renewal and the Sacred Work of Grief*

CHAPTER THREE

SLOW DOWN YOUR CRAZY CHILD

Remembering

May your day be filled with a rare honesty and a fierce grace.

May you unwrap your pain, remember your sorrow.

May you blanket sorrow by weaving threads of compassion and patience.

May you return and lovingly care for your child.

May you find love, acceptance, others and nature to nourish you.

May you remember you are profoundly precious.

May the soles of your feet be rooted as cedar tree.

May you question everything, connecting to your original way.

May you learn delight in the unknown.

May you trust the winds will return you, or guide you home.

May you slow down, settle, and breathe in the wonder you once were.

May you pause to root. Patiently and courageously rise and roar.

May you observe that hardship is temporary or lifelong, bravely walking beside life.

May you slow down, listen and dance with the harmony of your heart.

—*Written and Lived by Dawn King*©

* * *

Throughout our lifetimes, we will experience disrupted nervous systems. Life can take its toll, and turbulent times are the price and the underrated power for living this gift of life. Many of us have habitual operating manuals and stay in the sustainable pace of high stress jobs, working too many hours, financial struggles, relationship challenges, health problems, and various degrees of traumatic events and experiences. Our nervous systems can become unbalanced, and this is where there are great lessons to be learned in how we can care for them. Most of us have never learned to restore or reinstate a sense of calm or peace.

Slowing down by walking in nature, saying no to doing more, scheduling weekly self-care dates, playing a sport you love, listening to music, journaling or writing a poem, turning off your phone or background noise, or taking five deep breaths in through your nose and out through your mouth can be incredibly soothing medicines.

Whether these are learned behaviours of the past or the build-ups of stress, poor sleep, mood disorders, or repetitive frustrations, one of the miraculous ways to manage our imbalances is to care for them regularly so we do not default or relapse to old ways. The emotional or behavioural build ups add fuel to the fire. Slowing down can pacify the accelerated pace of trauma responses.

Candace was a thirty-five-year-old single mom who came to see me for social anxiety and depression. After history-taking, Candace shared that since she was nineteen years old, she had always been in unfulfilling relationships. She wanted to have a traditional relationship path of marriage and children but the partners she had were either unfaithful, unavailable emotionally, worked too much, or abused substances and simply avoided deeper communication or willingness to change or improve the relationship. When Candace came to see me, she had just become single again and was searching desperately for another partner. It had not occurred to her that slowing down and practicing being out of an intimate relationship may be a solid solution. In her past, it also became apparent that her parents' divorce when she was ten years old had a huge impact on her sense of safety. Her mother was a very passive person who did not stand up to her father's top-down approach laced with emotional and verbal violence.

Candace learned to become afraid of expressing her thoughts and feelings openly and honestly. Consequently, she never had a relationship with a man who could, either. The slowing down practices and principles allowed her to start having an intimate relationship with herself first. She stayed the course of creating a connection and understanding who she really was. After one year with no partners, her solo and sacred journey developed her into a confident and assertive young woman. Within two years, she was married to an emotionally available man. Ten years later she remains in a beautiful marriage with two children, anxiety and depression free.

Countless clients I've worked with over decades learned passive behaviours as well. The seduction early in a relationship can quickly turn to a mistake based on seeking security rather than slowing the entire relationship process down. Instead, it's important for an individual to take inventory as to the health of the person they're dating, to let themselves grow, and adjust to slowing down the speed of commitment. My father was violent physically, so I made sure to marry for safety, not a mature or available love from a healthy person.

Neuroscience tells us our brain is hardwired to look for danger and stress in order to protect us—essentially, it is always looking for ways for cues and control. Internal sensations or feelings such as fear, sadness, anger, or anxiety repeat consciously and unconsciously. External experiences such as work, home, or social environments create this as well. If we are dysregulated or unable to properly manage emotional responses, all these messages can give the signal we are in harm's way and have us living in a constant state of survival and stimulation, or dysregulation.

What we could understand and then practice is that the danger is not entirely real. If we exposed ourselves to more calculated risks, we would be able to train this part of our brain and behaviour. We can feel fear and protect ourselves, and absolutely must do this in real danger. Learning to discern what is real and what is imagined or constructed from the past or the future takes new teaching and unlearning certain actions and practices. Unfortunately, this trains us to be stressed or absorbed in what we cannot control, such as our internal or external environments, creating an over-dominant fight or flight state. When this happens, we habitually respond with a stressful high alert mind and body. As children or adults, we likely did not learn how to create calm or peace when we needed it. It is never too late to relive our childhoods or at least take full accountability for our pain and pleasure.

Most of us are raised this way for generations. When I ask clients how their family handled conflict, clients will answer in one of three ways:

1. "We didn't." (Avoidance)
2. "We hit or yelled." (Aggressive)
3. "We never talked or felt anything; it was never discussed." (Passive)

Slow Down Principles

- Patience—give me patience quickly. Most of us would prefer results come faster and discomfort would go away quickly. However, slowing down creates such a connection that the need for speed changes. Practice pausing before you speak. Begin to listen to your inner voice and hear what he or she has to say. Begin to accept people's decisions whether you agree with them or not. Focus on yourself.
- Purpose—often this is chased or constructed as a "to-do" list or the elusive "what or who am I?" Purpose can be as simple as learning to calm down or connect to ourselves and others.

This is a richer, more meaningful direction—what a great purpose to slow down, the next step is usually revealed.
- Practice Powerlessness—what can I control? My thoughts, feelings, and behaviours. Trying to control others or every outcome is tense and often leads to inner turmoil. Practice staying in your own lane. Faith in the unknown will improve with time and practice.
- Play—our creative minds and our physical bodies need to move to regulate and resource emotions and activities that bring us joy. Sample activities include: coloring, dancing, drawing, trying out for a play, singing a song, writing a poem, playing a board game, laughing often, riding on a swing, taking a polar bear swim…whatever brings us joy and laughter.

We often don't take enough risks, and our barometer for uncertainty or the unknown becomes our prison as we hide behind or feel trapped in a fearful cage. Taking risks is both an inside and outside job. Emotional or physical isolation shackles our nervous systems. Self-reliance can be a horrible entrapment for trauma or dysregulation. Self-reliance is lonely and limiting. We need other people in our lives.

Risk for you could mean expressing how you honestly feel, applying for that new job, committing to something or someone, writing a book chapter, starting a new business, making a new friend, joining a social group, and countless other actions you may be putting off.

No matter what we do or don't do we are taking a risk. Risk is on a continuum, and even the smallest step can strengthen us for the next one.

One client of mine had difficulty calling her doctor because she feared they would find something, so she avoided the risk. Of course, doing so increased the risk that she needed medical care and perpetuated her anxious feelings. Asking for help was her remedy, and she invited a friend to hold her accountable for making the call within three days or that friend would come over and sit with her while she phoned. I suggested she ask her to go to the doctor with her, alleviating her isolation and moving through her uncertainty and fear. We all can and will lose our way. Sometimes we need a companion to walk alongside us. Taking too big of a leap can be too much. A single step is all that is required. Slow down, and ask for help if you need it.

The Myth of More

When was a time you remember life slowed down enough that you felt peaceful, grounded, and/or calm, and it lasted? I recall this was when my sixteen-year-old cat suddenly became gravely ill. My daughter rushed him to the vet hospital, and it was determined he had advanced kidney disease. I was able to bring him home because I was willing to be his caregiver. IV fluids, meds, and constant warmth and snuggles were required as I established his care somewhere between palliative and hospice. The end of life can have an abrupt but meaningful opportunity in teaching us to slow down. It can take us from overdrive to first gear in a heartbeat.

The next ten days I was able to move all business and clients to an online platform and I delegated all that needed to be done to slow down and take care of him. Animals and nature have taught me that the circle of life constantly changes and is unknown. So, when my cat's diagnosis required the care for his departure, my life slowed down. It snowed every day. It was -25 degrees, and my cat and I snuggled in my room under the covers. We reminisced, I cried, and I asked him if he would be seeing my golden retriever, Milo (also my cat's dear friend who had passed away two years prior). I asked him if we were going to meet again and constantly checked in to assure his physical comfort was at the forefront. The traditional medicine wheel of indigenous teachings shares that winter is a grief cycle, or a slow-down cycle, as I like to think of it. For ten days, I slowed everything down until my cat was euthanized by the vet who cared for his passing in the comfort of those warm winter blankets.

Slowing down is always an elixir for anxiety or trauma. Anxiety is fast in pace and usually thrives in the future or races to the past looking to convince itself there is something of which to be afraid or forewarned. Unresolved trauma is imprisoned anxiety. Anxiety is our learned behaviour thinking we can control fear, hesitancy, or apprehension of the unknown. Yet the more we speed things up and the more we avoid uncomfortable emotions or situations, the more anxiety is created. Slowing down is common sense. It is simple but not always easy. Our need for speed and control is ever-increasing alongside mental health concerns, mood disorders, insomnia, high blood pressure, anxiety disorders, addictions, expressing to ourselves or others that we never have enough time, lack of purpose or direction, chronic fatigue, or obsessive worry about what may happen.

The more we avoid discomfort internally or externally, the more we solidify the belief that the lion is going to eat us. None of us would have ever learned to walk or to avoid a hot stove if we didn't get in touch with pain or uncomfortable emotions. But we are so busy running away from boredom or avoiding fear by seeking reward instead of strengthening our capacity for uncertainty and adversity. We certainly understand the value of stressing our physical bodies to exercise and build muscles and cardiovascular endurance, but emotionally and mentally we habitually avoid discomfort.

Anxiety is fueled by attempting to control the external world, yet its resolution is within. It is a learned behaviour in a culture that endorses it.

The Lost Art of Connection

Connection to self becomes the elixir, the grounded and rooted inside job. Connection to self and others has become a lost art and practice. The volumes of anxiety, addictions, depression, suicide, chronic illness, stress, burnout, and "the myth of more" invites us to take off the adhesive bandages, masks, and badges of doing and return to the roots of identifying a sense of connection to ourselves first.

The reality is our external world is never enough. More time and focus need to be spent within ourselves. We live in a culture that consumes and focuses outwardly. This behavioural habit and mindset perpetuates the do-aholic or a more-aholic, and fractures the foundation of self-worth in our children, families, and our emotional and mental health. This fast pace has led to a performance-based, self-worth culture, where more is better.

When is enough enough? Let's slow it down so we can remember what is already here. Growth and goals rooted in our sense of meaning and purpose are different pathways than the myth of more. We ought to practice rooting and connecting to reverence, grace, and the courage to create, making our relationships or environments a better place for all.

By slowing down, we catch up to life, and it leads to co-creation and accountability. Slowing down awakens honesty and opens self-reflection.

Our addictions to pursuing more and other addictive behaviours can be made complex, but the idea is simple: to avoid pain or to gain pleasure and repeat. We have become a pleasure-seeking culture forgetting that discomfort and emotional adversity is a real necessity. Industries have been born from our more-aholic and do-aholic dopamine desires, and products sell in the billions of dollars. As this industry seduces us into more happiness, more money, more perfection, more performance, more fixing, and more advice giving, more facade is created and the further we disconnect or realize our real selves. If we just live in the moment, change our stories, mindset, or location, everything in our world will all of a sudden become better… but it never does, so we buy it again. How much more do we really need? What would life be like if we slowed down enough to deeply understand what we actually want or need?

Based on keeping the dopamine levels high, positive thinking, many coaching programs, workshops, and well-intended self-help books give us instant or temporary gratification. Always seeking more without actually finding or seeing the benefits leads us to believe the answers are outside of ourselves. Happiness and a positive mindset are for sale because pain does not sell. We suffer needlessly and silently because we all have emotional pain being held in the secrecy of our depression, anxiety, or overachieving. Our do-aholic, workaholic, and/or more-aholic lives cycle incessantly.

More of anything is also rewarded by our culture. Achievement breeds the need for accolades and a false sense of self. Perhaps, we learn the more we do or achieve, the more people will love us or approve of us. Maybe then we will be or have enough. If we only do, more people will approve of us. It definitely feels wonderful and likely better than how we currently feel, but it is only temporary. Today the "Badge of Busy" is a dysfunctional measurement of self-esteem and worth.

Moreaholism is also normalized. As I currently look out a downtown office window, I see a liquor store on one corner and a marijuana store on the other. The ingestion of these substances for many of us can become primary strategies for escaping pain or stress, seeking to feel feelings of pleasure and motivation and or avoiding emotional discomfort. Then I look at a billboard that has designer clothing seducing images of the ideal body, along with the products they represent as things to consume. When could we ever be enough if we are always comparing or competing with others as to the quality of our house, family, appearance, or bank account? But it's precisely what marketing intends you to do. The hypnosis that feeds the pace and perfectionism is, "Buy more to make yourself feel better."

We have created cultures and people who are performance and perfectionism-based, seeking approval from others versus self-acceptance. Many of my clients are entrepreneurs, corporate managers, health care professionals, teachers, professors or retired individuals who are recovering from perfectionism and performance-based moreaholism. They have controlled people, companies, money, and external results but have lost control of themselves and their identities, feeling empty and purposeless. Addicted or obsessed with glossing over feelings, they share how they live as perfect little robots self-judging and comparing themselves and others while giving 110 percent in the pursuit of earning the approval of others. These are the people who are one person at work and go home and are a totally different person. One client who owned a multimillion-dollar company and retired at fifty-five asked me in a session, "How can I successfully run a company but cannot be here for my family or fix all my problems?" One client called it her fake facade life. What a horrible life cycling around which coat or mask to wear in order to adapt to the environment. It's another reason why "more" is for sale. It lets people jump to the next hit of dopamine. As the Eagles sing, "Life in the fast lane, surely makes you lose your mind." Truth.

It takes at least a few months (not thirty days, one week, or twenty-four hours as many self-help "experts" claim in their marketing campaigns for their programs) for my clients to find their way to a sustainable speed of self-acceptance, imperfection, compassion for self, and a curiosity versus a perfection to always get it right and control everything. But first of all, they have to muster up courage and be vulnerable to see what is not working. A radical form of honesty and self-acceptance is the result, and every aspect of their life can begin to slow down, calm down, and connect to themselves and begin to thaw the real in the feel and the flow of everything they cannot control. What we could use more of is actually the ability to deal with discomfort. How we can do this is to slow it all down and begin to feel our emotions instead of always chasing or forcing a false sense of control or getting rid of anything uncomfortable, only to replace it repetitively with something or someone we become addicted to and dependent on.

We have countless examples in the performance-based world of athletes, celebrities, and business who have benefitted from therapy. Eventually the "myth of more" had them seeking professional help because it caught them impulsively acting out, destructively doing or saying something, or ending up in rehabilitation or trauma therapy.

We became an ultra-protective controlling culture, believing that doing everything for our kids (young and old) would make it easier for them and for a short moment of pleasure, we would feel better about ourselves. This belief and those actions are denial and delusions at its finest. An example of this would be a sixty-five-year-old parent with a thirty-six-year-old son who is addicted to heroin. To make a better life for her grandchildren, she pays for her son's car payment and gives him a thousand dollars per month for groceries. She defends her actions, believing he buys groceries with the money.

Or take the father who made his son a business partner in the family business in order to protect his son from cocaine and alcohol addiction. The father sincerely believes that if he keeps his son busy enough at work, the addiction will simply go away. The son rarely shows up for work. He draws a monthly salary, and when he does sober up for a few days, the father has the family home ready for his son to come back, take a soak in the hot tub, raid the fridge, and sleep in, only to relapse within three days because vodka is as easily delivered as it is as accessible at the drive-through. Rinse and repeat.

Maybe you can relate to having people doing too many things for you or perhaps you do too much for others. As a culture, we want so much to avoid discomfort that we have created a "Dopamine Nation." In her book, *Dopamine Nation: Finding Balance in the Age of Indulgence,* Dr. Anna Lembke writes that the developed world is full of nations whose people have avoided regulating and learning about emotional homeostasis by using substances or behaviours to stay in the pain or pleasure centers of the brain.

One of the reasons we do not slow down or know the benefits of doing so is because we have learned to do for others what they need to do for themselves. Maybe we jump all over an opportunity to tell someone what to do. Do you need to be in control of everything or everyone? Maybe you are the only one who can "get the job done" or nobody else can do it as well as you can. We can fixate on other people's problems while denying our own. These behaviours as coping mechanisms worked as a child trying to adapt and create some form of safety. If we just people-pleased or lied or made someone or the environment feel better when all hell broke loose, it all made sense to do these self-negating behaviours to meet our temporary emotional needs.

Codependency and enabling are key phrases at which people cringe or don't understand. I simply define these behaviours as doing more for others (assume, worry, fix, advise, etc.) than we are doing for ourselves so we can avoid accountability and keep a false sense of control.

The problem with these behaviours is they can prevent us from slowing down because we are constantly trying to control others…and that is stressful! What happens is we take the care away from people and eventually learn to be passive aggressive, not standing up for our own needs or even knowing how. Here are some behaviours that cycle the dependency on other people to make our "bad" feelings go away temporarily until they return moments or hours later:

- Doing what is not our responsibility.
- Not saying what we need.
- Putting the needs of others first, placing them before our own.
- Doing what other people are capable of doing for themselves—and need to do for themselves.
- Meeting people's needs without them asking for our help.
- Getting involved in what isn't our business.
- Doing more than our share when someone asks us to help.
- Forcing our help onto people when they don't ask for it.
- Taking care of other people's feelings or problems; neglecting our own.
- Facing people's consequences so they can avoid them.
- Speaking for people and not letting them speak for themselves.
- Making excuses for others.

- Spending much time assuming others' thoughts and feelings and adjusting our own based on them.
- Trying to solve others' problems by fixing or giving advice.

The Difference Between Codependency and Self-Responsibility

These behaviours cross the boundaries of others because we assume we know what is best for them. Also, we honestly do not trust people to take care of themselves. This creates more anxiety for us because we avoid dealing with the discomfort and distract ourselves by focusing more on others. Notice how often you try and "fix" others' problems and how it actually makes you feel. The problem we are fixing is likely not our own, and the other person may not have even asked us for help. That's an actual problem we need to care for within ourselves. Even as a therapist, I have a total and complete belief that the person I can help has the ability within themselves to do so. No one is broken or needs fixing. An experienced guide, a healthier perspective, a great listener, or a trusted safe person to be with—yes!

"Give a man/woman a fish, and you feed him/her for a day. Teach her/ him how to fish, and you feed her/ him for a lifetime" is a quote from the Chinese philosopher Lao Tzu, and it certainly rings true.

The slowing down and noticing these addictions and habits creates a sense of calm and relief. We can begin to practice the concept that we only know what is best for ourselves. We feel incredibly relieved of responsibility when we allow others to manage themselves, therefore reducing anxiety and triggers.

It may be a worthy challenge to practice waiting to be asked for help or advice. This way your focus can be on your own challenges, and it likely will leave more open space for your solutions and resolutions to those problems. You let yourself in to care for your overall well-being, which may likely be past due.

My client, Clare, came to see me for depression. Her history revealed her father and mother were rigid with rules, as well as emotionally unavailable. She expressed that her childhood home was where we "didn't talk about feeling anything." Also, Clare shared she was unhappy in her sixteen-year marriage. She was unhappy not because of any "bad" or "abusive" reasons. "He's a good man," she told me.

Women will often minimize, justify, or validate their passive behaviours or needs by saying things like, "I am unhappy, but he is a good person because he works and he takes care of our son." It can become a defensive way to shame and guilt ourselves instead of getting real and honest that we want more out of the relationship. Over a few months of therapy Clare became aware of how codependency in the relationship was the root cause of both of them being unhappy. When they first married, Clare was in a deep and long life-threatening time of depression. Her husband was understandably fearful and did everything in his power to run the house, earn the money, and pay the mortgage.

Over the course of a few years, Clare came out of her depression, but the relationship was now based on fear and control for both of them. If Clare took a risk by trying something new, her husband would doubt her and hold her back, believing that taking a new job or trying a new sport (healthy risk) might make the depression return. He had become controlling over her every behaviour, and she submissively never said what she wanted or needed because it would come with words of doubt and fear. For sixteen years, words of encouragement and empowerment never came—the depression was always the problem, and her world remained scary, fearful, and passive. As long as Clare played small, her husband could feel safer and in control of his uncomfortable feelings.

Clare spent three months unlearning her behaviours based on the enabling and codependency relationship traits. She and her husband sought individual therapy to understand the origins of their codependency, and eventually I saw them collectively as a couple. Once Clare began to assert her own needs, her husband began to reduce his fear responses.. Eventually, over time, both began to believe more in each other and were less fearful of the depression. She has had no symptoms of depression for seven years, and their marriage grew them both up, emotionally and behaviourally.

Canadian Gary Robbins has been a highly disciplined and driven runner all of his life. He even became an accomplished and consistent Barkley Marathon invitee. COVID restrictions, however, taught him to slow down. Robbins wrote on his social media platform what changed for him and how COVID taught him to slow down. He reflected on his time with family and the commitment time required for his extraordinary competitive running career. His son who was five years old would soon be ten. Robbins said he did not want to run for eight to twelve hours a day, missing his son's childhood. So, he announced he was going to continue to train but for shorter distances. This form of self-reflection and inner inquiry is often found in slowing down. Gary was able to connect to deeper and truer values and accept that achieving or accomplishing specific meaningful goals of the past had changed for him.

The consequences of hurrying are present in every aspect of our health. Insomnia, anxiety, depression, high blood pressure, chronic pain...the list is endless and obvious.

The benefits, however, of slowing down offer hope and inspiration for the change. As we deny, dodge, and distract, we do it at a pace that is not calming or connected. It is externally driven, yet internally we are depleted. Slowing down creates connection to self, others, and environments, and reels us back into what we can control versus what we cannot.

One of my goals of this book is to share how you can regulate, reconnect, and renew to become your own medicine. My hope is this book reconnects you to the work that was never passed on to you but is within you.

We need mentorship; we need to be taught how to use what is within us. In my opinion these skills could be taught as life skills equal to learning how to swim, drive, read, write, finance, and communicate. How many unhealthy relationships have all of us experienced? Rarely did we have role models readily available to teach us the road to changing or creating healthy ones.

Why Decelerating Helps

Deceleration has its benefits. By slowing down, we can feel the emotions we are experiencing and describe them. In doing so, we can understand them and let them help lead us to a healthy response. For example, clients will say they don't have anger or feeling anger is inappropriate, yet anger can be an insightful emotion. If we slow down enough to listen to our anger, it can be the catalyst to teaching us to prevent future pain or it can fuel us to take action. Slowing down helps us channel emotions into actions that serve us well. Anger can move us into action.

Another benefit to slowing down is we begin to learn we are actually unproductive when we force things to happen or remain emotionally agitated because we keep trying to fix people. Slowing down helps us focus on feeling what is truly important. Slowing down actually takes us further but is not performance based; it is human-centered.

Slowing down allows more intentional awareness of making decisions. Impulse control is a problem in our culture, yet slowing down expands creative thinking and provides a space and place within us of making and planning for possibilities.

Doing more doesn't always equal doing our best. How much of the landscape do we see at 120 kilometers per hour? While constantly talking or on our phone while walking, how much life do we take in or see? While working eighteen hours a day, how much connection to our loved ones do we experience? Or, while having the family at home with the TV screen on, how much distraction or connection can we have? Savor and slow down—or, as my son says in his cooking methods, "low and slow."

Pushing our physical limits can be anaerobic; slowing down and constantly and methodically taking the next right step is sustained by moderate intensity, low-impact movements, which lower the risk of disease. Slowing down emotionally, physically, and/or mentally will provide for self-reflection, sincere connection to others, and a realistic day-to-day life we can walk alongside rather than chase or be chased.

Slowing down is an internal and external experience. We can breathe, exhale, journal, draw, write, or create in any way. When we slow down, we can become better listeners to ourselves and to others.

We can slow down our eating, make eye contact, observe people or nature, be with someone dying or sick, or simply help a person in need of our presence.

Accepting our limitations is often frowned upon in this "more is better" world. What actions can we take to support the slowing down process? Saying "no" more times than we say "yes."

When asked how you are, how often do you answer "busy" and actually feel the tension and stress in your body? Life in the fast lane will surely make you lose your mind.

I don't think we want to get older faster. I think we could all benefit from marinating in the emotional experiences of our lives. Not stay stuck in any of them but to listen, to learn from all of them. So many people have imparted their wisdom on death and dying. This is where it really slows down. They do not tell us we will want more of anything. Their wisdom and words share we will only remember how it feels to have lived and never how fast it took to finish.

As I say farewell to Yang cat, aka Mr. Miyagi, the winter outside my window insulates my reverence for his life in mine. I will miss his nightly warmth at my head on my bed pillow. As he aged, his wheezing pacified my occasional nighttime loneliness. Similar to my golden retriever, Milo, Yang was always there for me. Animals are another way to invite and invoke the slow-down sensations of a charged nervous system. Animals and/or our pets never rush life or abandon each other—animals tend to share and walk alongside us in return for food, water, shelter, safety, and chin rubs. Grief never dies, for love is never lost. These past ten days have slowed me down in honour and respect for my dear companion. Goodbye for now, Yang Cat. I do hope we meet, snuggle, and purr together again.

It is my hope that by slowing down you reconnect to your heart and the heart of others. It is here the lost art of connection is found, remembered, and savored. Slow down, and stay there for a while.

Ashton
R3 Program Graduate

What a remarkable year it has been; job promotion, R3 graduate, and ultramarathon finisher!

Three years sober and an entire year more wonderful than the last. I'm so thankful for the beautiful connections I made in R3. I love having these amazing women to journey with me. I am incredibly grateful for my renewed connection to self and nature. I am blessed to have Dawn in my life as having a powerful female role model has provided me the kindling I need to ignite my fire.

I've been in my new job for nearly an entire year and what a journey of professional discovery it has been. I feel a sense of pride when I reflect on the magnitude of improvement I've experienced in setting boundaries and communicating congruently. I am glad to report that I am staying in my lane and not taking on what isn't mine – emotionally and professionally.

This year has been remarkable. I faced fear, shame and pain and fuelled my personal growth and change with passion, self-compassion and sometimes anger. With significant support from my mentor and R3 sisters, I travelled back to my roots and gently explored the misunderstandings and experiences of my past where the seed of my defenses were planted.

Knowing my roots allows me to know myself and if I know and care for myself – I am always prepared. Most importantly, I no longer say "yes" when I mean "no." I am in charge of me and responsible for my growth and happiness. What I am giving myself today, will be passed down to my daughter tomorrow.

I feel pride every day knowing that my daughter will have a different experience than I. I am a cycle breaker. The chains of codependency have loosened and my over-responsibility is diminishing with each experience. I have a deep sense of freedom knowing that I am responsible for putting my own needs first.

As the New Year begins, I gladly and with open arms, accept the uncertainty that the future holds. I will remain rooted and will grow with faith and love towards the light of my future.

RISE

"For me, singing sad songs has a way of healing a situation. It gets the hurt out in the open into the light and out of the darkness."

—Reba McEntire

"Music acts like a magic key, to which the most tightly closed heart opens."

—Maria von Trapp

CHAPTER FOUR

THE SECRET LIFE OF THE SOMA: SINGING

Why I Sing

I sing to feel
To release brain and body bondage
I sing to connect
To unlearn the disconnect
I sing to ignite
My home fire, my heart
I sing to grieve
Oh, that little girl
I sing to breathe
Releasing her pain and tensions on the exhale
I sing to nourish
Song is medicine, melodically moving into the known and unknown
I sing to find my way home.
Lived and Written by Dawn King ©

* * *

As a child raised in immense uncertainty and volatility, I believe my innate survival skills were gifts in disguise. Nature was my original therapist and a way I could attach to some form of grounding within and gain a source of belonging. But nature also taught me to sing.

When I was five years old, I would sit and sing with my K-tel record player. I only had one vinyl 45 record which played a song by Gilbert O'Sullivan titled, "Clair." While the madness of a tumultuous household repeated daily, singing was a way to enter a world of emotional possibility. In this world, I could feel joy and sense tensions ungluing from my bones. My body relaxed, thoughts stopped racing, and my outer world became a far-off land. Singing created an environment I could control within me and an emotional freedom where my brain and body emerged into a calm I could count on.

My parents used to play an 8-track tape of Glenn Campbell and Anne Murray in the car. I would sit and revel in the boat-like motion of the back seat of their 1968 Chrysler Imperial and sing every word of all the songs. Singing was a form of teletransportation that took me to a safe emotional planet far into my own imaginary galaxy and prepared me for the return to the family home environment.

By the time I was five, my mother and father were both in the throes of alcoholism. Hearing music and singing was a somatic medicine, a way for my body to release, a welcomed soother, and healthy escapism. I didn't know this at the time, but singing was a healthy, dissociative way to take my emotions and give them some place to regulate and relieve themselves.

When my parents were married, they would be in the living room entrenched in wild parties, and I would avoid the sounds by being alone in my room with music. Once they divorced, I stayed with my mother for a short period of time, but due to the effects of alcoholism, she was unable to provide a cure for my sleepless nights and excessive restlessness. I could never wake her up when I was afraid at night, so I crept into her bedroom, curled into a ball on the floor, and sang myself lullabies. I hummed until the sadness and fear of being alone seeped from my body, and exhaustion finally forced me to surrender. In my distressed slumber, while lying on my mother's purple shag carpet, I grasped and sang for some sense of safety and attachment.

Within months, I moved to my father's apartment with his wife number three. I stayed with him until I was twelve years old, through wives number two, three, and four. My father was a ruthless con man, always running from bad business deals and angry women. We moved twenty-two times in two years so he could stay ahead of his lies and embezzlements. My mother had abandoned herself and me to severe alcoholism by this point and was nowhere to be found. While my father kept running and conning his way (with me in tow) through three provinces—British Columbia, Alberta, and the Northwest Territories—the radio playing in the cab of my father's truck was the airwave to my own world of joy.

By this time, I had picked up a larger repertoire of songs, while he picked up women and slept with them in the camper attached to the truck. I stayed in the truck's cab and sang while pulled over on the side of the road on remote highways of the Northwest Territories. I would sing to songs of the time by Roberta Flack, Olivia Newton-John, The Carpenters, Elton John, Steve Miller Band, John Denver, BTO, Chicago, Anne

Murray, Dr. Hook, Gladys Knight and the Pips, Steely Dan and Bette Midler, Doobie Brothers, Barbara Streisand, Earth Wind and Fire, and Linda Ronstadt. They provided free somatic therapy during this stressful and tumultuous time in my life. The melodies shifted and transformed my physiology, thoughts, feelings and bodily sensations of despair and isolation into one of wonder and toe-tapping emotional freedom.

Throughout my elementary school years, singing continued to soothe my ruminative thinking and temporarily settled fears of the past and fears that raced into the future. I struggled in school, isolated and submissive, and never attached to anything or anyone except art, music, and poetry. My first singing gig was when I was nine years old. I sang, "Let's Go Fly a Kite," a song from *Mary Poppins*, to the parents of my grade five class.

School provided a temporary solace from an alcoholic and abusive household. My stepmother at the time found her emotional relief through beating me. Every day after school, I could count on my head banging into walls, the stings of face slapping, and the painful tug of hair dragging. But I could also count on listening to the caws of the crows, the screams of the ravens, and the tweet of the black capped chickadee, which miraculously sings its own name. At night, I listened to the low and high notes of the snowy owls, and even the ptarmigans' glottal call calmed my wounded body. Their songs returned me to my voice. There I could escape from pain. I was a scrawny little kid who feared the adults I lived with. I rarely talked to anyone, but the birds and I sang until a sleepy, safe slumber overcame me.

My father's Travelling Trauma Roadshow came to an end when his wife at the time pointed a shotgun at us both. With the shotgun in her arms, she raged through the house looking for my dad. I bolted for the closet, calling to the northern lights to take me away. Hours later, my dad pulled me from the closet while fighting off my stepmother. I had no idea where the gun went, but I was shoved into the cab of the pickup truck with a strange woman sitting in the middle of the truck cab. There was always another woman, and my sensories and hyper-radar locked into who she was at that moment. I started to sing a song inside my head. It was "Let Me Be There" by Olivia Newton-John. Her voice was so resonating and warm to listen to that singing in my head and humming into my body allowed me to escape the despairing miles ahead.

After thousands of miles of singing in my head in the "getaway car" helping us escape, yet again, from my father's delusions and deceptions, I was dropped off at my mother's apartment in Vancouver, British Columbia, Canada. I hadn't seen her in two years. By then my mother had made herself very sick with alcoholism, I'm certain as a way to avoid the feelings and memories of her traumatic past. I was now in the huge city, away from nature and in the franticness of a one-bedroom apartment I shared with my mom in the final stages of alcoholism. I was to attend a school three blocks away, but school was never a place for a little girl like me. After twenty-two moves, I'd left every school I was ever in within weeks of arriving. By the time I was eleven years old, I had a natural instinct to look for danger and adult manipulation in everyone and every environment. I could not add, read, concentrate, engage, or connect in the classroom or in social settings. I was in a constant state of numbness, claustrophobia, and anxiety. School was a psychological prison because I couldn't move—mentally, emotionally, or physically. I was always hyper alert for the next move and had a total sense of disconnection to anyone or anything.

I quit school at the age of thirteen and worked two jobs. I resorted to lying about my age in order to land a cash job washing hair at a hair salon and washing dishes and cooking at a local golf course. At the same time, I enrolled in barber school—a trade I quickly learned to love.

When I was fourteen years old, my mom moved from our apartment to go live with her boyfriend. I was thrilled. For the first time in my life, I had no interference or destructive influence from adults. I was independent and on my own. My two jobs allowed me to pay for rent and food. I had no friends or family in my life, and for me, this was the perfect place to be… alone and safe… me and my songs. Work was now my new escape hatch. I learned that I could count on no one, and I built a secret and safe life as a minor living on my own. With this lifestyle and arrangement, I could stay clear of the pain and chaos of living with adults.

I had money to spend from working two jobs, and I knew the one constant relief I could feel was through singing. I took the bus to Radio Shack—a very popular music store at the time—and I bought my first Pioneer turntable, mixing board, speakers, and a Shure microphone. I was fourteen years old and free—a teenager with her music, emotional, and mental freedom! Every moment I wasn't working, I would sing in my apartment to my imaginary world. It was the only friend I knew or needed.

Looking back and understanding trauma, I now understand how singing saved me. I sang because the grief of losing my childhood and my family was overwhelming and debilitating.

To hold grief, sadness, abandonment, abuse, and aloneness is too much for anyone to bear. I needed to sing for the voiceless child within me. Singing regulated my emotional brain, moved my traumatized body, and calmed my anxious thoughts. In contrast, music can also remind the memory to return or react to a derogatory time in our lives. We can settle and be transformed into beauty by choosing music and song that uplifts our emotional tides.

Had I not found singing as a soothing and emotionally uplifting form of escapism, I know I would not have lived beyond sixteen years old. As an isolated child on her own, independent, withdrawn, and medically undiagnosed with trauma symptoms, I would have found a destructive outlet instead of singing. Singing was and remains one of my most powerful medicines.

When we sing, we release tension. Typically, the defences and psychological limitations guarding us will also reveal themselves when we begin to sing. This is why most people sing alone or after too much consumption of drugs or alcohol. These substances are notorious for lowering inhibition and affecting judgment. We are afraid of what others will think or have been criticized into believing we can't relax and release our voices in self-expression. Our self-judgment and the limiting beliefs others project onto us can keep our voices in the closet or limited to the shower, but singing along to an emotionally moving song can heal us as well.

When I became a mother, singing was always in my home. My oldest son enjoyed singing and playing guitar as a young child. He had a close school friend who repeatedly said, "Stop singing, you sound horrible." The first time I heard his friend say this in our home, I asked him respectfully to stop. For several years my son was aware his pitch was all over the place, but his desire to sing and play guitar kept him moving forward, and he formed a band at eleven years old. The same friend also joined the young band and kept up the criticism. I continued to encourage my son to sing. From the ages of five to fifteen he sang for the pure joy and expression of how it made him feel. Like me, I believe it helped him move through his own childhood adversity. At sixteen years old, after years of refusing to listen or believe his friend who criticized him, my son was invited by the local symphony company to join their tour as a paid tenor!

My son is now a father of two, and as I write this, my granddaughter is starting piano lessons. He has given his love and appreciation of music to the next generation. My other children share their older brother's affinity for music, and it is a bond that connects them synergistically. To this day I can count on my second son to always have an epic playlist on in the background of his home. Now in his thirties, we still share in singing to his vast variety of setlists while making traditional meals together. His knowledge and repertoire of music is extraordinary, and he is quite the dancer. My daughter was in the high school jazz band and played a stellar sax. We spend a lot of time together in the car driving and singing, enjoying her setlist of a remarkable music repertoire. This mom is blessed to have musical, creative, expressive adult children, and song is a joy we all share together.

Our body is a beautiful instrument. We house sounds in every cell of it. For over thirty years, I have invited clients to sing to connect to the emotions they have minimized or shackled, and/or to ignite and restore their joy and exuberance. Most of us were raised being told our feelings were good or bad, and this creates a separation to the flow of emotions. We cannot connect the dots and bouquet of feelings if our knowledge about emotions is so black and white. Like music and sound, we cannot find a harmonious connection if we do not understand how to create it.

Emotions are similar to nature as feelings are cyclical, but only if we can connect and move them into meaning. Emotions or feelings will arrive as they move through us and return like the waves and winds. Many emotions create an anxious feeling, causing us to be unsettled, afraid, apprehensive or uncertain, but we have put all our emotions into separate boxes. While anxiety and depression can be debilitating, our resources seem so limited to change or improve. Medications, therapy, courses, retreats, and lifestyle may or may not help or have any lasting effects. Depression and anxiety are an engineering of numerous emotions. Although not always true, for the most part, depression lives in our unresolved past, and anxiety lives in fears of the future.

Prior to feelings of anxiousness, depression, or any limiting or stuck behaviours, we may not be aware of the variety of emotions that stack themselves or tangle and entwine our realties. Sadness, frustration, grief, anger, uncertainty, uneasiness, fear, and so much more are emotional patterns resulting in constructive or destructive outcomes. Destructively living in our emotions can create a sense of being stuck or not able to get out of them. It creates the emotional or mental abyss, such as in a constant state of anger, depression, grief, sadness, shame, and guilt. All these emotions have meaning and purpose, as we discussed in the "Emotional Medicines" chapter, but if we do not move from the emotional abyss, we are likely hurting ourselves or others unnecessarily. We are taught to blame our behaviours on anxiety or depression, but singing lets us feel what is underneath and possibly dislodge the patterns. It is real in the feel. Emotional fluency and connection can be brought back to life through singing because it allows our emotions and body to move together. The flower cannot blossom without losing its petals and returning itself back into the earth, and singing and emotions naturally unearth or renew our self-expressions by allowing us to grow into our own emotional safety.

We self-medicate through alcohol, drugs, prescription drugs, and/or we use behaviours such as reading books, Netflix binging, phone scrolling, social media, and other distractions to avoid feelings that stack behaviours and symptoms of anxiety or depression. Emotional escapism is often our way to do this. Hours stack into days or years as we seek emotional relief or stimulation through external means, all to avoid uncomfortable emotions when we could be singing softly and somatically through song. Singing can soothe our pain. Through singing, people can express or release embodied trauma and connect to the feelings that were never heard or cared for.

If you seek to identify the numerous emotions that create patterns of anxiety and depression, within those emotions you will find the medicine to heal them. Perhaps through singing you will feel your grief, loss, or gratefulness. Singing can give your pain and pleasure his or her voice back by allowing the defences created by the walls of trauma to fall.

When you are watching a performer sing a song, how does it make you feel? Singing improves your mood, and it can also connect you to deep, unexpressed, or remembered emotions.

We can all go back in time when we hear a song that reminds us vividly and emotionally of the past, whether it is a happy time or a very difficult one.

Singing releases the same feel-good brain chemicals as sex and chocolate—those emotional endorphins that create uplifting feelings. It is very effective as a stress reliever and can even improve sleep. Singing is a natural antidepressant and can calm our nervous systems. Singing changes the brain by moving vibrations (as in most instruments) through us, altering both our physical and emotional landscapes.

According to Heart Research UK, a leading United Kingdom Heart Health Organization, "Singing engages the brain's reward system. Research has now found that when we sing to any song that is on our mind, the act of singing aloud is beneficial to our health. In addition, singing helps improve the aging process." If you are a parent, maybe now you can identify why lullabies have purpose and meaning.

For the clients I see who experience disturbing memories, depression, or anxiety, I invite them to find a beautiful and empowering song or two and move to the rhythm while sharing their voices on the refrain. They often report that singing completely helps them out of despair or fear.

When overcoming or moving through anxiety our one-on-one therapy I often suggest singing as a joyful way to gain confidence and build self-esteem. Join a choir, try out for a musical theater part, take acting or voice lessons. Any time we take a risk to overcome the emotional shackles of fearing what other people think (including ourselves), we build up the psychological muscles needed to begin believing in ourselves.

A fifty-year-old client who was unable to sit for any length of time because he felt such restlessness and nervousness chose a timeless Frank Sinatra song, "My Way." The exercise I gave him was to start singing when he felt the onset of restlessness and to sing the song out loud, as if no one was watching. My client returned to his daily living and put this practice into action. After daily singing for only a few moments each time, he was able to teach his body and brain to relax. This singing assignment changed the unconscious patterns and involuntary anxiety responses he'd carried for many decades and allowed him to feel calm and peaceful again.

Singing releases stored muscle tension and decreases the levels of a stress hormone called cortisol in your blood. In addition, many Alzheimer's and dementia researchers and caregivers use singing as a way to restore memory in patients. High levels of stress or toxic stress combined with emotional or mental defences can have a person overthinking, overanalyzing, and ruminating over the past or future. Many of my clients share concerns about their poor and noticeably declining memory. By talking or singing about their feelings, they can, for the first time, connect emotional patterns that are creating anxiety and, therefore, affecting their memory. Once these defences are defined and singing is introduced and repeated daily as a way to calm, clients notice anxiety levels reduce considerably. Sleep and memory improve as well, as it is all connected.

Singing moves muscles and emotions. It is not coincidental when working with individuals and trauma victims to discover that loss and grief are often the precursors for anxiety, depression, and/or the increase of addictive behaviours or substances.

A woman who had experienced the death of her mother fifteen years prior came to see me because she was having social anxiety and panic attacks. She also shared that she had increased her alcohol intake and used marijuana to sleep.

After several sessions, I asked her to choose a song. She chose her mom's favorite song. Within the safe and supportive environment of my office, she mustered up her courage and finally began to thaw the grief she had been holding for fifteen years. As she sang, she moved through her tears, her profound sadness, and remembered her mom. In my piece of writing titled, "7 Lessons I Learned from a Red-Tailed Hawk," I write, "Grief never dies, because love is never lost." She had never truly made an effort to remember her mom. Yes, in thoughts or fleeting memories she came to mind, yet she entirely avoided feeling anything remotely uncomfortable. The emotions were too strong, and she was afraid to feel them. Singing gently released her pain but also restored the joy of remembering her mom.

Research confirms these personal experiences with music. Current findings indicate that music around sixty beats per minute can cause the brain to synchronize with the beat, causing alpha brain waves (frequencies from eight to fourteen hertz or cycles per second). This alpha brainwave is present when we are relaxed and conscious. To induce sleep (a delta brainwave of five hertz), a person may need to devote at least forty-five minutes, in a relaxed position, listening to calming music. Researchers at Stanford University have said that "listening or singing to music seems to be able to change brain functioning to the same extent as medication." They noted that music is something that almost anybody can access, which makes it an easy stress-reduction tool.

Singing moves emotions, brings forth memories, and recognizes and reveres our feelings. Singing is as primal as we are. It is simply natural, soothing our nerves and elevating our spirits.

An eighty-two-year-old brave and beautiful client came to see me for depression and anxiety. She expressed concern about her memory which she noted had changed in the past two years. I asked her when she stopped singing and listening to music. As it turned out, she had stopped singing and going to the symphony when her husband died two years prior. After weekly sessions of singing and grieving, her joy has been restored and her memory has improved. This now vibrant woman remembers where her car keys are and the door code to her condominium. She joined a local community choir three months later, free of depression, anxiety, and a noticeable change in memory.

Every life experience is stored in our bodies and brains. Let us sing to them and about them!

"Stillness moves me, movement stills me."

—*Dawn King*

CHAPTER FIVE

MINDFUL OF MOVEFULNESS

Why I Swim

I carry weight.

I inherited the cells of traumatic lifetimes.

I swim the open waters to free the debris and cleanse the dead skins of my survivalist ancestors.

The cool waters soothe the inflammations of the historically invisible wounds.

The waters know every scar and unburden them within a weightless womb.

It is here the heaviness is insubstantial.

I swim to feel the delicacy and connection to creation.

I swim to free the future. To not deny the pain but brave its cold discomfort, moving to it and through a kelp entangled forest

I swim captured in effervescent while evaporating grievances. My arms as the oars, are strong as spruce wood as bodily tensions exhale.

I swim to free the inheritance. I swim to leave ripples of grace and courage for those seeking to transform their pain into love.

—*Written and Lived by Dawn King©*

* * *

All my life I have felt relief and strength by moving my body. Swimming was my first love. Until I was nineteen years old I lived on the west coast of British Columbia where lakes and the ocean were as abundant as forests. My joy for the open water even had me representing Canada in the world championship triathlon for my age category which eventually reignited my joy of open water long distance swimming. Movement for me was never about competition, it was about release of tension. Many "races" I came in last or well behind any of the "winners". I am not a fast athlete, but I will finish. I have completed distances of 750 meters to 10-kilometer swims. I am at home in the water.

As a kid I had dreams of everything from being an Olympic athlete, to an emergency doctor, to a country vet, and a professional singer. When I look back, one thing these vocations had in common was a tremendous amount of physical movement. I think this was the release of mental and emotional suppression I could not identify then but the physical movement seemed right. It was a release, a calming balm on every level. In addition to swimming, I enjoyed skiing, lacrosse, and trail running as a kid. Movement helped me escape from the tumultuous family environments l lived in. At ten years old, I would leave in the middle of the night to snowshoe to the Hay River, and nobody knew or cared I was gone. This was the only way or place I knew where I felt calm…there I could breathe.

I would swim the lakes and oceans of the west coast whenever I needed an escape, which was often. The wildness of the Pacific Ocean pacified the anxiousness inside my body, and lake swimming immersed my emotions in a safety net I couldn't find anywhere or with anyone else. I am sure if nature and movefulness had not guided me, I would have turned to destructive behaviours and harmful self-medication. Instead, nature and movefulness were my healthy outlets as I quit school, worked two jobs, paid my rent, financed my tuition for barber school, and remained hidden from social services.

My need for physical movement provided relief from what would be defined as anxiety in today's day and age. I think we live in the age of anxiety, and anxiety is commonly used as a description, yet deeply misunderstood as to how to relieve it. There is an inner restlessness that is difficult to describe to someone who has not experienced it. I used to be described as a very high energy person, but determining high energy, anxiety, and a perpetual pattern of running from ourselves is highly misunderstood. That anxiety helped me run three marathons, complete twenty-two triathlons, and guided me to become an accomplished open water long distance swimmer. None of these athletic aspirations were for competition; all were for the love of how moving made me feel. All that anxiety and those dormant emotional volcanoes had to go somewhere! Moving created a channel for feelings—I didn't need to swim the English Channel!

I remember the first time I started running. It was horrible, heavy, and I hated it. I've always been a person who packs extra body weight with every traumatic experience. Through therapy, I was able to accept the fact that this bodily armour of fat has learned to protect me, but packing it around by running it off is hard work. As I built my endurance one kilometer at a time, as heavy and sluggish as I felt by the thirty-minute mark, it was as if I found second gear, and the flow I felt after the heaviness subsided was liberating. It was a groove. I also experienced many runs, where I spontaneously had to stop in a puddle of tears not because I was in pain, but involuntarily sadness overcame me. Because of the movement, I felt grief for the first time. I can now understand how moving my body guided me to emotional and physical release. I have sat on many forest floors and country roads heaving with sobs and releasing emotions that simply took over and asked to be let out. Until I experienced these spontaneous emotional releases, I had restrained everything within every cell of my body. Movement opened the doors and lightened my loads.

An inspiring story and book titled, *Depression Hates a Moving Target* is a memoir and an inspiring story of Nita Sweeney, who was forty-nine years old, overweight, and suffering from crippling depression of bipolar disorder when she caught the running bug. Sitting on her sofa in her pajamas one weekday, she saw a social media post about a middle-aged friend who had taken up running. She leashed up her yellow Labrador retriever, Morgan, and headed out with a kitchen timer in hand, running for only sixty seconds. She kept running a little longer each day until, two and a half years later, she crossed the finish line at the Columbus Marathon. Nita Sweeney became a runner.

Nita has since completed one ultramarathon, three full marathons, twenty-nine half marathons, and more than a hundred shorter races—and through it all has faced her many fears, learned to cope with her bipolar symptoms, and, through the power of movefulness, discovered inner strength she didn't know she possessed.

I think all of us are aware of the benefits of movement, and for me in particular, open water swimming returned honour, care, reverence, and respect to me and my traumatic past. Running released strong pent up and painful emotions, and open water returned me to the waves of my courage and pure joy. I am fortunate to live in an area of British Columbia, Canada, where we are surrounded by pristine lakes and ocean. Within five minutes, I can be in the open water, swimming, watching eagles overhead and schools of fish below me, and enjoying mountains surrounding me with their forests of fresh air. Relief and release are the best way I can describe how movefulness exhales bodily and emotional tensions. It's gentle, there is no forcing anything, and it's as if the wisdom of every experience of my life knows what it needs to remember or release.

I trust my body to show and share what it needs, and this takes practice through movefulness.

It can be quite frightening for some who are not yet comfortable in the water to swim with waves and winds. For me, the power of nature is an experience to merge into rather than tackle. Feeling the power of nature reminds me of the courage, strength, and connection to resilience of moving through traumatic events. Rather than buck the tides, I move through them.

As a child, I could not control my environment. Like so many of us, there was no one to talk to who could explain or share that all emotions were meaningful and what to do with them. No one was parenting, asking what I needed, or providing any sense of calm or safety. So as the feral child, I was gifted with what nature provides in her companionship—a natural rhythm for emotional movement. Nature self-regulates her harmonious systems. She doesn't need our help, and each and every part is ecologically intelligent.

McMindfulness... Over 99 Billion Served

In recent years, a tsunami of mindset, mindfulness, and meditation strategies and practices have been touted as the go-to or cure-all for trauma and for the anxious or busy brain. The common wisdom has been widely shared that if you do not feel peaceful or happy, you should meditate. But when I tried to meditate, I felt an unsettled, progressive build up of energy, which created agitation and a sense of feeling trapped. All that emotional charge needed to move, not sit and be still.

My clients also share the same experience of having high volumes of pent-up, shackled energy in their bodies and did not know how to release the tensions into an internal sense of overstimulation. Clients will share they cannot sit still, that inside of them is a frenetic feeling they cannot still or calm. I used to call it Mach 5, which is faster than the speed of sound and measured typically through supersonic airplanes at 3,800 miles per hour.

When we feel this sensation inside our bodies telling us to calm down, settle down, relax, or insisting we change a belief or go meditate more, it is the equivalent to keeping a horse in a stall with no availability to move or release tension. Imagine any living being trapped in the confines and cages of captivity, without any freedom to move. Movefulness is necessary for the release of bodily tensions and naturally helps regulate emotions.

There have been enough studies to show that meditation is not the cure-all for an anxious, busy mind or the symptoms of traumatic responses. In fact, it could bring on further emotional distress and even psychosis to those who have a history of abuse. During meditation, there is the danger of painful repressed memories spontaneously surfacing. According to Dr. Willoughby Britton, who has studied the adverse effects of contemplative practices, meditation can lead people to some dark places, triggering trauma, or leaving people feeling disoriented. She released a study that identified fifty-nine different kinds of negative meditation experiences, and her research has also shown these distressing experiences are not limited to people who have a history of mental illness.

Clients have also told me of unfortunate experiences where they have gone on seven- or ten-day meditation retreats where there is limited food, very little sleep, and/or a rigid schedule of sitting for hours in discomfort and distress, only to be told to work through the pain. Within four days, clients share they had to call family or friends to pick them up and take them to the nearest emergency department where they were diagnosed with symptoms of psychosis and anxiety attacks.

As Harvard University's Center for the Developing Child report states, the term "ACEs" is an acronym for Adverse Childhood Experiences. It originated in a ground-breaking study conducted in 1995 by the Centers for Disease Control and the Kaiser Permanente health care organization in California. In that study, "ACEs" referred to three specific kinds of adversity children faced in the home environment— various forms of physical and emotional abuse, neglect, and household dysfunction. The key findings of dozens of studies using the original ACEs data are: (1) ACEs are quite common, even among a middle-class population. More than two-thirds of the population report experiencing at least one ACE, and nearly a quarter have experienced three or more. (2) There is a powerful, persistent correlation between the more ACEs experienced and the greater the chance of poor outcomes later in life, including dramatically increased risk of heart disease, diabetes, obesity, depression, substance abuse, smoking, poor academic achievement, time out of work, and early death.

The ACEs report also discovered that meditation could aggravate trauma, and they recommended movement and action as the elixirs for trauma symptoms.

When one has suffered from trauma, one's mind is filled with anger, rage, hatred, self-loathing, and other turbulent emotions. Negative thoughts spin their mind into turmoil as they want to lash out, scream, and get even. Intrusive thoughts fill their days and nights, and forcing their thoughts to stay still and silent is often just not possible.

In asking someone with trauma to pay close, sustained attention to their internal experience, we invite them into contact with traumatic stimuli—thoughts, images, memories, and physical sensations that may relate to a traumatic experience. This can aggravate and intensify symptoms of traumatic stress, in some cases even lead to retraumatization—a relapse into an intensely traumatized state.

Movefulness is a practical self-awareness process which integrates physical and emotional self-awareness. I see many people moving in gyms or outdoors but with headsets, music blaring, and or non-stop talking to their training partner. There is nothing wrong with this—enjoy the tunes and the friendship. Movefulness encourages the awareness that we are connected to our bodies and emotions. We are tuned in to our senses, needing to feel the energy of emotions move through us and out of us in order to find calmness or regulation.

While mindfulness or meditation practices can indeed help people improve sleep, reduce rumination, or decrease anxiety, there is a common thread with my clients that it also does not. Stillness or even rhythmically breathing may not work; in fact, they can aggravate and annoy. We need to create more fluidity and larger motions in order to get the survival brain into the frontal lobes. Our fight and flight responses are millions of years old, and they don't always want to tiptoe out of the cave. The emotional or physiological storage of our life experiences are largely unconscious. Movefulness can provide some space for the emotions, and in many cases, involuntary responses to fear or past traumas give us a chance to change the response to it.

We can find this through singing, dancing, creating large bodily movements, or stretching and discovering what form of movement feels best for us. Some people feel a calmness through yoga, while others run, ride horses, play sports, work out at the gym, hike, swim, or walk the dog. It is a worthy and highly therapeutic self-discovery process to determine what "feels" best for the release of our tensions. For us trauma responders, if we do not move it, we lose it! Movefulness, in fact, stills us within.

In my group's programs, we move our bodies by walking in nature, painting, drawing, writing, and singing rhythmic associations. I have not met one client in thirty years who says, "I can't wait to go meditate." For us trauma responders, it makes us feel like imprisoned animals, but in nature we are free to feel the ground beneath us, maybe for the first time in our lives.

For a time, a nurse came to see me for her anxiety. She was unable to sit still and was diagnosed with ADHD. She wanted to be able to sit in her lovely backyard and relax without fidgeting or having to "do something". Her body and brain had learned after years of childhood adversity that chaos was more comfortable than calm. She had never learned to be still, let alone be comfortable with stillness. Her thoughts raced while sitting on her sun deck in her lovely backyard. Inside she felt "busy" and unsettled, so naturally she wanted to do something with the discomfort, and drinking an alcoholic beverage had always worked, temporarily.

For twenty years, happy hour was her temporary relief to medicate the discomfort of the anxious or chaotic sensations she felt in her thoughts and body. Alcohol temporarily stopped the racing thoughts and calmed her bodily tensions. After attending our outpatient treatment program for alcohol dependency and not having any alcohol for four months, her cravings returned with a vengeance, and as hard as she tried to relax, she could not. Once her brain had removed the alcohol, we could then begin to support the deeper reasons as to why she drank. Alcohol temporarily relaxed her, but she would not stop at one or two drinks. Her daily intake was six to twenty-six drinks, and for over twenty years, she was a "highly functioning drinker" who held down her job, but her anxiety worsened and so did her alcohol consumption. Her alcohol use started at sixteen years old as a way to make her feel good and temporarily escape a volatile household. She had never learned how to calm her body or her thoughts, so she overachieved and never slowed down. When she did, she returned to the anxiety and nervousness.

I slowly introduced humming as a way to create movefulness to this client. The recommended exercise was to sit in her backyard chair on her deck, and when she felt nervous sensations arise, she was to hum her favorite song only for a minute or so. After one week, her craving decreased by fifty percent.

The following week, I suggested she sing out loud this favorite song and begin moving her body to the song's rhythm, tap her foot, get out of her chair, and begin large flowing movements, as in dancing. She looked at me, with embarrassment, and said, "Really? I'd feel ridiculous (feelings of embarrassment and self-judgment were trapped) but okay, I trust you. This humming thing is working, and I'd like to get rid of these cravings. I'm afraid of relapsing." After another week of humming in her chair for a minute, followed by tapping her toes, she reluctantly felt the embarrassment of being in her backyard solo dancing and singing out loud. Thirty seconds built into three minutes of physical and emotional liberation, and the feelings of fear, shame, and embarrassment transformed into joy, playfulness, and fun.

Now, two years later, this client has never had to fear a craving, and the restlessness in her thoughts and body have never returned. She learned that chaos was more comfortable than calmness, and through movefulness, she could teach her body and brain that calmness was in her control. She had learned self-regulation through movefulness.

Next time you feel restless, unsettled, nervous, anxious, or chaotic, remember a song that uplifts or creates a calm feeling within you. Begin humming it sitting or standing. Start moving your body with it. Now, sing out loud the words and move your body to the melody. You'll feel like a rock star and may advance to an air guitar or microphone in no time.

A more subtle movefuness exercise I teach my clients is to use the five senses to simply name self-soothing images on the fingers of their hand.

You can try it now or when you feel anxious, nervous, or overstimulated.

1. Notice and number how anxious or unsettled you may be feeling (Scale: 0 not at all - 5 highly unsettled).
2. Look at one of your hands and start to imagine or bring a sense of curiosity to what makes you feel calm.

   - Sight - Move your thumb and imagine an image that is calming to you. Examples such as seeing yourself sitting or lying on a beach or swaying in a hammock, looking out to an ocean or lake.
   - Touch - Now wiggle your index finger sensing what feels calming to you such as the softness of a flannel sheet, stroking your pet's fur, a baby's skin, or a warm blanket.

- Sound - Next, move your middle finger and think of a sound that is calming, like a bird singing, piano music, waves on a beach, the wind in the trees, or a song.
- Taste - Move your calmness ideas over to your fourth finger. What is a taste that is calming? Maybe the taste of chicken noodle soup, a sip of tea, or hot chocolate?
- Smell - On your small and last finger wiggle what is an aroma that is a calming smell? Perhaps it's the smell of lavender, maybe the ocean air, a rose, chamomile tea, or a freshly brewed cup of coffee.

3. Notice now what number you were at the beginning of this exercise or how your level of calmness has improved. Has your unsettledness or anxiousness decreased? Go back and repeat until you are more comfortable or to a zero in sensation.

The vast and rarely recognized behavioural pattern we suffer from is the avoidance of inner pain. Trauma responses, defence mechanisms, uncomfortable emotions, and unprocessed emotions all make up that which is unfelt and unheard in our stories.

As we avoid our inner emotional pain, we lie in the depths of despair, depression, anxiety, addiction, or various methods of self-harming. Our nervous system is held hostage as we attempt to cope in chaos, calm with substances, remain in unhealthy relationships or move quickly to the next one, feel with food, fix with sex, drown with drugs, alleviate with alcohol, or get on the warp speed of doing and the upper of overachieving.

Avoidance: The Power, the Pleasure, the Pain

Yes, we can recover, but some do not or cannot. As a recovered adult child of abuse, neglect, and abandonment, it has always been human connection that has healed me deeply. Therapy can be highly effective, but sometimes it isn't. Social medicine has and always will be an underrated antidote to help us in all ways. Another person's love, listening, acceptance, presence, connection, that stays with an individual long enough, has the potential to move most of us from the shadows and can give encouragement to reach over and open the curtain to a lighter way. Movefulness can be experienced in countless forms of self or social expression, such as talking, socializing, walking, or even attending a concert.

I encourage all of us not to give up on ourselves or others we love without serious regard and reflection. Asking for help is usually the most difficult, but often the most impactful first step. And know this: If you seek professional help, it may take many attempts to find the person who actually can and does help. I was the only one in my family of origin who sought ways to change, understand, and pave a new path for the inherited destructive behaviours and beliefs of the generations before me. The two marriages I attempted failed mostly because of my ineffective past coping mechanisms and theirs as well.

Too many people don't work on their part. In fact, they just leave and repeat the pattern. Relationships are difficult, and we bring with us our past pain. But we can also bring with us and into our relationships the resolution of that pain and the hard work it takes to grow a relationship. People who are unwilling to change simply find a new relationship or isolate in aloneness. History can and does repeat itself without changing the cycles and actions within them. And yes, of course, many relationships are not healthy to remain in, and it is time to sever the chords. My life experience and thirty years of observation is that people leave far too soon because they give up, period. Yes, some relationships will not change, and for your own well-being, it is time to go your separate ways.

Movefulness connects our relationship with self and others. An internal relationship, such as with anxiousness, trauma, and low self-worth, or an intimate relationship or friendship that is stagnant or complacent needs to move. It is as if we are adding water to a plant. If we nourish it, the plant will grow. Without movement or action, a new approach, new resources, or growth, there is a slow and steady death toll. We all need to move emotionally, physically, mentally, and spiritually.

Isabel
R3 Program Graduate Letter to Self

You are an amazing, courageous woman of infinite worth and thoughtful action. On one hand, I can't believe it, and on the other: of freaking course you did.

I am so incredibly, amazingly, overwhelmingly proud of you. In one year, you have transformed your life. It's almost unrecognizable. But of course, it was you. It was always you.

You began to take your wishes and desires, your freedom seriously. You began to trust that support, possibility, that abundance was really there for you to step into. You accepted what came to you, you took all that in and walked forward into the incredible unknown.

And here you are. A bit re-invented, and also completely and utterly you. Solid, more sure than ever before. Sure in your worth, permission to live your worth. To take your dreams and desires seriously. To advocate for yourself and your needs. You got clarity. You got support, you got incredible results. Me and R3 for the win!

Your lovely new home, your changing relationship with your son, your incredible R3 community, your new interests and activities. The way you are supporting and nurturing yourself. You are welcoming new people into your life. New joys and the experiences you have been yearning for.

You have transformed your financial reality. You are financially independent. You achieved your biggest, scariest goal. And you stood up and took action for your happiness.

Isabel, you are incredible, magnificent. Look at you resting, enjoying, living, savouring. I am so incredibly proud of you.

Dawn and her R3 program has taught you to cross the thresholds and come into a new phase of life. You are an adult. You are caring for yourself in ways you never dared or bothered to in the past.

You have softened into your emotions with courage and compassion. For yourself and for others around you. You are transforming your relationship with your son and have courageously said "No" to entanglement and drama. I see you making choices daily for your wellbeing and peace.

I remember years and years ago when you were ready to put down the sword. Well, my darling, you are actually doing it. The most courageous and beautiful thing you have ever done.

I see you smiling at yourself in pleasure, and touching yourself with respect, care and kindness.

You are a superstar. A flower. A cedar. I am achingly proud of you, and grateful to you and soo very, very happy for you.

With love, adoration and YES!!

ROAR

"If you bring forth what is within you, what you bring forth will save you. If you do not bring forth what is within you, what you do not bring forth will destroy you."

—*Jesus of Nazareth*

CHAPTER SIX

COURAGEOUS COMMUNICATION

Road to Recovery

I do not celebrate a recovery day.

We are all in recovery, restoration or rejuvenation: the gift of life

I celebrate the truth about trauma, this we all share.

I celebrate my loving adult children and a small group of loyal friends who are empathetic and collaborative in caring for me. The brave but few who will sit with you, until the light and their love, thaws the protective armour formed by betrayal, abandonment, stigma, shame and fear.

We are not flawed or broken or in need of fixing, we are burdened and shackled in unexpressed pain.

Avoiding pain, the shadows, creates more pain.

It is big business to not be real, to dodge, deny and positively fake our pain and our path.

Buyers beware.

I celebrate the bold and beautiful ones willing to fall, fail and no longer fake anything.

I celebrate the courage and the radical realness required for an authentic life.

I celebrate you, Brave One. I Celebrate You.

—Written and Lived by Dawn King ©

* * *

The amount of young people struggling with low self-esteem and self-worth seems high. Many health professionals and parents tell me young people are struggling with depression, process addictions (shopping/social media/relationships/gambling addictions), and various degrees of anxiety problems.

A global report launched by The Body Shop in March 2022 identified a crisis for young women around the world, with one in two women feeling more self-doubt than self-love, and 60 percent wishing they had more respect for themselves. I am incredibly thankful for the parents, physicians, teachers, and health professionals who are insightful and see our health and our behaviours are connected. Our healthcare systems care for our physical bodies, but our emotional and physiological pathways are all deserving of the understanding of connectedness.

Although I see both men and women in my practice with varying degrees of struggle with emotional isolation and self-love, I am over the moon elated that I have the opportunity to support these young women who seek higher self-esteem and self-confidence. It is tough being a human, let alone a young person in school or headed out into the world on their own.

In her honest, transformative, and essential book, *If Women Rose Rooted*, psychologist Sharon Blackie calls this withdrawal and isolation "The Wasteland", wherein lies our despair and emotional withdrawal. Dr. Blackie encourages us to sweep away the veil, to understand what is broken and what needs to change. She shares what it means to live in "The Wasteland".

> *"The wasteland is the hollowness within us, for we are reflections of the hollow world in which we live. To embrace it might mean that we spend our lives doing work or relationships we hate or dislike in order to feel secure defining ourselves by that work which we're paid to do for others, and yet find ourselves wondering why our hearts are breaking. It might mean we wrap ourselves tightly in busyness and noise and never have enough time and anxiety and panic and wonder why eventually our bodies break."*

Social isolation is not social medicine. We need each other, and we are required to rise from deeper roots which allow our true security; it is hard work and a necessary threshold and rite of passage. Life itself will share with us when we are off course, so we root, heal, reflect, restore, and renew. We prepare to rise again. Roaring our unique pathways is a stage in which we attach to a truth within us based on our inside passage, not what the external world believes we should be doing. Our voices, including our inner and outer communication, are congruently and assertively expressed when:

> *"We rise rooted as trees and we may need to travel a long distance before we find the new skin that fits us and before we can learn to be comfortable in it—but first we have to commit ourselves to the journey. We have to awaken from our torpor, commit to life instead of the desiccated half-life of the Wasteland. We must shake off the false skins we're cloaked ourselves in, we must let the old die to make room for the new, before we can discover who we are really meant to be and what our work in the world is."*

I see so many young men and women in the wasteland as they transition from high school to university, marriage, children, and the social stages for which we endorse and award a badge of approval. Most of our young women are not prepared for these transitions, as most of our mothers have done exactly the same. We can get stuck in the wasteland and stay in a passive behavioural pattern as generations before.

When we are in the depths of despair, it is often an accumulative effect of unexpressed emotions. We may have become passive or stayed in the shadows or retreated far too long not speaking about what is honest and real for ourselves. Teaching ourselves and encouraging others to roar—to express themselves openly and honestly—requires assertive communication...or as I like to call it, courageous communication.

We have a significant disconnection in our healthcare system. There is emotional difficulty for the client or patient to share their struggles or be transparent about their mental health and/or addictions with a medical professional. Our medical systems do not support a doctor or health care provider the time to sit down and actually connect, let alone understand emotional and mental imbalances or addiction. We have to take the time to allow a patient or client to safely open the door to their pain. Once again, life in the fast lane of health care leaves both patients and healthcare professionals without the time or awareness that connection can be the greatest gift to creating treatment and compassionate care. We experience the same disconnect in the counseling therapy world. Well-intended healthcare professionals help many people with depression or anxiety, yet are simply uninformed or unaware of the amounts or types of addictions people engage in behind the scenes.

I know approximately 60 percent of the men who enter our outpatient program for alcoholism have gone through a sleep clinic to assess sleep apnea. Not one assessment or healthcare provider has asked about alcohol use or other substances. Too many people end up in an emergency for addiction-related illnesses with no treatment. They go home with medications, detox, and sadly repeat the same cycles which can be unnecessarily deadly. I see the same challenge in treatment centers: people arrive at a rehabilitation or addiction treatment center, and they return home thirty to fifty days later with no support or plan for continued recovery but living alone, trying to cope without the addiction. The continuum of care is tragically lacking. Courageous communication will help improve this for all of us personally and professionally.

There are three characteristic styles of how humans verbally communicate: passive, aggressive, and assertive.

One hundred percent of the people I see struggle with being fully honest, fear communicating, or are left not feeling listened to in relationships. Most relationships are challenged because the communication is based in passive or aggressive communication—either not saying anything at all or saying things they don't mean to say (but maybe they do and have been afraid to) that can be hurtful to others. Passive or aggressive communication is a volcano either way. It is a slow, steady build of resentment, sadness, anger, and frustration waiting for the rumblings to seep out or fully implode into an aggressive screaming, hitting, throwing, supercharged release of pent-up emotions. Whether passive or aggressive, either communication style has side effects in our relationship to self and others. Passive or aggressive behaviour paralyzes us in analysis, self-loathing, self-pity, self-anger or hatred, self-harming, and/or acting out in words or actions that hurt others. We have all done this in various forms and degrees. The goal in roaring is to remain rooted and rise into an assertive and courageous style of communication.

Styles of Communication

Passive

Passive communication gives the responsibility to anyone and everyone who wants to feel in control except ourselves. It gives people permission to mistreat us and allows others to take responsibility and make the decisions. Are we in control of our finances, self-esteem, worth, health, careers, choices, and decisions? Yes! But passive behaviour can reduce feelings of self-worth because it gives the power of decisiveness and responsibility to someone else. Self-esteem is earned, and it is earned by taking risks and being courageously assertive in what we say and do.

Basic passive behaviour looks like this:

- Anticipates unpleasant consequences (assumes, fears, and futurizes)
- Withholds (says nothing and withdraws)
- Avoids and assumes conflict
- Says yes when we would rather say no
- Tells the problem to uninvolved people
- No eye contact
- Apologizes unnecessarily
- People-pleasing

The problems arise in communication when we are passive because there is no change. If I avoid or minimize a problem, there is still a problem. Passive behaviours also create tension in our muscles and bodies in general. If we find ourselves ruminating or overthinking, chances are we are passive and have unexpressed feelings, wants, or needs. This can lead to depression, guilt, and inner or outer self-punishment. We become irritable and hostile, and our behaviour will likely convert to aggression.

Aggressive

A person who behaves aggressively may dominate an individual or group or use humiliation or attacks to gain control. They often criticize others and can be intimidating with both body language and voice. Aggressive communication speaks loudly, interrupting or talking over others. The tone of voice is loud. They may also:

- Insult others' opinions and glare rather than making appropriate eye contact
- Use the word "YOU" in most sentences
- Threaten
- Generalize—use words like "never" and "always"
- Demand, command, and tell people what to do
- Blame others for mistakes
- Bring up the past and irrelevant history to leverage their righteousness

Aggressive communication is not considering other people's views or feelings. This may also take the form of ignoring people, rushing them, or bossing and belittling them.

Courageous Communication: Assertive

Basic human rights are lived and expressed in courageous communication, such as:

- Having dignity and self-respect
- Saying no when justified, without feeling guilty
- Expressing your feelings
- Asking for what you want directly
- Feeling good about yourself
- Being able to change your mind
- Negotiating and reaching compromises when conflict exists
- Being able to make mistakes

We have heard of or recognize these behaviours, but practicing and knowing how to implement them is where the change happens.

We all are aware that knowing something and doing it create two different results, and this is where courageous communication becomes our dear friend. Courageous communication has us walking, crawling, or running out of the wasteland.

Courageous, assertive communication looks like this:

- Making eye contact
- Moderate tone of voice
- Makes a statement of the situation
- Makes a request
- Expresses a feeling
- Speaks to the person involved

Communicating and acting assertively is an interpersonal skill that helps people to maintain healthy relationships, resolve interpersonal conflict, and prevent one's needs from being stifled or repressed.

The benefits of being assertive include:

- Gaining self-confidence and self-esteem
- Understanding and recognizing your feelings (reduce the simmer or erupting volcano)
- Earning respect from others (confidence is desirable)
- Improving communication (deepen connection/intimacy to self and others)
- Creating win-win situations
- Improving your decision-making skills
- Creating honest relationships
- Gaining more job/career satisfaction

Why Do We Avoid Communication?

Some people fear the repercussions of acting assertively or may lack the skills to express themselves effectively. We may believe we don't have the right to be honest and share how we truly feel. Far too many times we were told not to feel, think, or say what our true self-expressions actually were.

We often are in a relationship for safety and approval, but we may call it love. Most clients I see have a patterned history of always being in a relationship. From one trauma or safety net to the other, they keep depending on the relationship to make them feel better about themselves. So many of us have not grown up with any emotional fluency. We may marry for safety, or we may stay too long because the current relationship is better than the last one. We may limit our ability to give ourselves time to understand ourselves first and instead, hope each relationship will fill a void.

We can cause a disconnect with people we care for and/or those with which we are trying to build or restore a healthy relationship. Why or when can communication become ineffective?

Here are a few instances in which that may occur:

1. We were taught and validated to avoid discomfort. Think of the physical muscle vs an emotional muscle. Which one do you use more?
2. Assumption—"My feelings or opinions don't matter, and I am going to assume yours are based on everything in the past I have experienced with you."
3. Argue the facts instead of what matters (heart vs head)—"I feel hurt you talked about my business with a family

member or friend...." Instead, the person argues about what was said, not what was felt.

4. Give up far too soon by using defensiveness rather than openness and being uncomfortable.
5. Use exaggerated descriptions such as "you always," "you never," "every time," or "all the time." Examples: "You are always telling everyone our business," or "You never listen and always ignore me."
6. Shut down when it gets difficult—you may do this or the other person may do this. Maybe you distract by staying on the phone, or look the other way, or walk out of the room or shut the conversation down by saying something sarcastic or demeaning.
7. Accidently misrepresent or minimize your feelings. Rather than express a variety of feelings you may say, "I'm fine," "it's no big deal," or "whatever."
8. People-please to avoid fear of rejection, not being good enough, say it wrong, or having to get it perfect.
9. We are often poor listeners either distracting ourselves or already making up a narrative of what to say or advice to give.

Communicating courageously will not guarantee the other person will change his or her behaviour and give you what you want, but it will help you establish limits and boundaries with others.

Do you want to be right, which is a form of controlling at all costs (fear), or do you want to connect (freedom) with yourself and others in a courageous and more vulnerable way? Either we deepen our level of intimacy and resolve our relationship problems or we move toward finding healthier ones.

Courageous Communication Strategies

The purpose of these strategies is to improve the relationship with self and others. These strategies are intended to be implemented within healthy and safe relationships.

- Substitute trying to win with seeking and trying to understand.
- Substitute jumping to conclusions with actively listening.
- Substitute offering suggestions with asking questions.
- Substitute trying to control people with trusting them.
- Substitute projecting with patiently taking the time to reflect.
- Substitute trying to change that person and start accepting them.
- Substitute betraying yourself with setting healthy boundaries.
- Substitute offering unsolicited advice with offering support.
- Substitute blaming with taking responsibility for your actions.
- Substitute settling or minimizing your standards with asking for what you need.
- Substitute avoidance with being respectfully honest with your wants, needs, and feelings.

A fifteen-year-old student was referred to me because she was withdrawn in her school classes and spoke very little to anyone. Her parents and teachers felt she was isolated and withdrawn emotionally. We practiced the principles of courageous communication, and within six weeks, she had made significant changes. She was completing her homework assignments, she had found two healthy, new friends, and she renewed her sense of self-confidence while coming out of the slithery and demeaning shadows of "I don't belong, I am not worthy, and I am not enough."

Another young woman had gone to university, and after nine months of studies in a new city with no friends, the university asked her to return home because she had failed her courses and was terribly sad and depressed. Isolated in her room with nothing but screens or social media to keep her company, she felt she had let her family down by not giving her "best performance." She was self-loathing and abandoning any part of her that she once thought was of worth. It is incredibly sad to me that any of us are left alone and isolated, not knowing where or how to reach out for help.

Or another behaviour I think most of us can relate to is being or knowing a person who travels around looking for love, affection or approval only to fall short or miss the mark of having a relationship with a healthier person. Try a new approach. Try being on your own for a while and spend time getting to know yourself. When you start really enjoying your own company, it may be a sign that you have given the time and self-awareness to who you are and what type of relationships are healthy for you. Trauma usually looks for trauma. Codependency often looks for problems and a person to rescue. Take a step back and let time and new approaches heal.

There is a common misconception that single means lonely. Lonely can be another uncomfortable emotion, but here is the point. How do we deal with our loneliness? Some people have learned to find another relationship, filling that escapism with temporary comfort but maybe in an unhealthy relationship, again. Most couples I see are already lonely, and they have been together for anywhere from five to thirty years. This is sad and not necessary. Chronic stress from relationships can be traumatic.

The ability to find healthy people and activities to care for your emotions is less likely to have you land in relationships that are based on dependency. This is where a relationship can add to your already solid sense of self-awareness and self-worth.

Lisa is a thirty-two-year-old woman with two children from her first marriage. She came to see me because she was concerned about her unhappiness with the relationship she had been in for five years. While sharing some history, Lisa described that both she and her partner were obsessed with what the other was doing and with whom. Jealousy and mistrust created even more verbal firestorms. They both used sex as a way to make it all better, but only temporarily. This young woman came to realize after a number of insightful therapy sessions that she had involved herself with relationships that had a surface intimacy. This particular relationship provided physical intimacy but lacked emotional intimacy. Both Lisa and her partner were emotionally guarded where neither was vulnerable with emotions, and they both had poor listening skills.

It takes a few unsuccessful relationships or one brave and beautiful one that is willing to improve or change to figure this all out. Every relationship brings hurt, trauma, or insecurity from the past. A healthy relationship can bring any history of pain and provide a safe and vibrant environment to realize a deeper sense of intimacy based on trust and vulnerability.

Creating healthy relationships involves understanding our past traumas and feelings and learning to deal with our emotions, in part by sharpening our communication skills.

You will know you are in a healthy relationship with yourself or others when:

1. You feel a solid sense of self-worth, security, and respect in a healthy relationship.
2. You have a desire to act, improve, and try something new. Empowerment occurs in healthy relationships; you feel like taking action toward growth.
3. You naturally feel a vitality and increased energy. When you contribute to the relationship and you receive it back, there is movement. If you feel drained or exhausted in any relationship, it's time for a change within it.
4. You feel a broader sense of self-awareness. Because you are in a healthy relationship, you see parts of yourself that you may have not seen or acknowledged.
5. You feel a strong sense of intimate connection in a healthy relationship. The more connected you feel, you usually have a stronger desire to have friends and connect more with others outside of the intimate relationship.

Effective and open communication is fundamental in quality relationships. A common belief in our own psychology is often that we shouldn't have problems, but they are inevitable and help us grow. Resolving problems can be accomplished with effective, honest, open, and intentional communication.

We can resolve conflicts, and differences are to be expected, but these differences can be challenging. Stronger emotions can arise, and returning to the courageous communication skills will become more natural.

Healthy relationships help us heal and feel, and they also help us sustain our trauma recovery and sense of well-being. A common quality of healthy relationships is common values. It's not as if we are taught to figure this out when we first meet. If there is one problem consistent in healthy relationships versus unhealthy ones, it is the ability and willingness to deal with change. Rigidity causes relationships to smolder and eventually die out when one or both partners continually refuse to compromise. The ability and willingness to change is essential, as the relationship needs of each partner may also change. The ability to deal with change is crucial.

We have a very strong and compassionate counseling/coaching team at my private practice, and at our Friday team meetings, our hope and most common statement is, "If he/she would be open to changing, this is all going to work out."

Human connection and communication are the essence of our fulfillment and our emotional and physical safety. Take a chance communicating assertively, and become more aware of actively listening to yourself and others.

"One of the most calming and powerful actions you can do to intervene in a stormy world is to stand up and show your soul. Soul on deck shines like gold in dark times. The light of the soul throws sparks, can send up flares, builds signal fires, causes proper matters to catch fire. To display the lantern of soul in shadowy times like these -- to be fierce and to show mercy toward others; both are acts of immense bravery and greatest necessity. Struggling souls catch light from other souls who are fully lit and willing to show it. If you would help to calm the tumult, this is one of the strongest things you can do.

There will always be times when you feel discouraged. I too have felt despair many times in my life, but I do not keep a chair for it. I will not entertain it. It is not allowed to eat from my plate. [...] In that spirit, I hope you will write this on your wall: When a great ship is in harbor and moored, it is safe, there can be no doubt. But that is not what great ships are built for."

—*Clarissa Pinkola Estes, Ph.D.,*
Jungian analyst, author,
and New York Times bestselling poet

CHAPTER SEVEN

ROAR REVERENCE

Adversity and emotional pain have taught me deep reverence. To define reverence is to describe respect… respecting ourselves, our pasts, our traumas, our talents, our limitations, our defeats, and our victories. Many of us did not live an ideal childhood passage. Even those who had that "ideal" childhood" have just as much loneliness or destructive behaviours as those who came from nothing—someone who receives everything can be left feeling as though they have nothing.

Recovering from our pasts ought to be the first thing we are taught to do as young adults. Remembering what the past taught us can help us with the wisdom to avoid or prevent those same mistakes. Most families have acquired generational trauma, and our thoughts, beliefs, values, rules, actions, habits, and behaviours were all modeled to us. And, it is all changeable. Rooting or settling into the best of the past and also the lingering regrets or limiting perceptions is great personal work. Rising involves remaining solidly connected to those roots while willing to step up, get real with healing those feelings of the past, and ground ourselves. In being brave and vulnerable, we become liberated, as well as set a new example for the next generation. Enough is enough. Mistakes were made. Misunderstandings from others are the guaranteed admission ticket into life. Remembering and revering every part of our past will strengthen our mental and emotional roots. What strengthens us and what within us remains fragile or unresolved? What has served you in the past whether it is traumatic or not? Self-awareness is rooting, rising, and roaring—the foundation for a solid sense of worth, trusting ourselves in the known and unknown. We can then congruently design and define our own unique adult ways into maturity, wisdom, creativity, and fulfillment.

If we utilize the suggestions in this book: Nature as a Healer, Singing, Movefulness, Slowing Down, Emotional Medicines, and Courageous Communication, each one of the chapters individually or collectively will lead to connecting our feelings of reverence for this life and all its limits and liberations. In comfort and growth, we create a natural reconnection to ourselves and ultimately others.

We can all have feelings of grief, anxiety, shame, and guilt, and we have used these emotions as shields of protection rather than appreciating or revering them with acceptance and wonderment. We somehow think something is wrong with us if we have these emotions, and we also have much confirmation as children taught either not to feel or to feel something other than discomfort. Big feelings in little bodies are scary, and far too often no one is there to help us move through them. We are left alone and unresolved. We all can think something is wrong, broken, or not working properly when we have anxiety symptoms such as fast heart rates, chest tightening, shallow breathing, or avoidance of anything or anyone that makes us feel uncomfortable, apprehensive, or afraid. We can be afraid of fear itself. I remember talking to a psychiatrist regarding a client, and she said, "I wish our patients with anxiety symptoms were taught how to work through fears or phobias that are learned, and then they would realize in their thinking that they are actually not going to die."

We are socially and culturally taught to have it all together. We talk more about celebrities and clothing than we do our feelings. We give more energy, attention, and admiration to the outside world (appearance, money, car, social media, rank, and how many "friends" we have). I strongly advocate for us to have reverence for our insides first—our thoughts, feelings, hopes, dreams, disappointments, mistakes, and victories—for all of it! The sheer amount of thinking and emotional hostage-taking we experience by seeking approval from others is actually a really good fear to have. It creates a healthy fear and self-awareness that we can use to keep our self-honesty, intuition, self-respect all in tack instead of the people-pleasing or perfectionism negating our own needs. Fear can also teach us what to avoid—a good fear to have!

Our fears can get shackled as well, because we are taught to think and believe that everything bottled up is best expressed by putting on a strong and fake front. Perfectionism is an agonizing and exhausting obsession. In my programs, I teach about how to overcome it. We believe we should be happy, calm, and "with it" all the time. Sadly, inside we are left alone in an emotional imprisonment of chaos, unhealthy relationships, overworking, worrying, meaningless mind-numbing distractions, or other destructive behaviours. It is difficult to have reverence for anything when we are overwhelmed with our stress or things that are not working out, and the uncertainties life can bring. Problems, however, can help us to grow, but trying to control everything, everyone, and all our perfect feelings is a life spent at a masquerade ball, avoiding our imperfections while dressed in the costume of a misunderstood protection.

How did we learn to avoid or stress out about problems and become perfectionists? Let's start with the masks we were taught to wear.

Masks From the Past

Swiss psychiatrist Carl G. Jung said, "We meet ourselves time and again in a thousand disguises on the path of life." The masks each of us wear on any given day, at any given moment, varies for all of us. We may have assumed the masks over a period of our life, or they may be an identity that was given to us. Families, caregivers, teachers, or cultures define identities for us such as athlete, princess, smart, dumb, abled, disabled, etc. These are all projected expectations of who or what others are or who they want us to be. This can all be very confusing as we put the masks on, either protecting those identities or trying to find our own and disagreeing with the expectations of others. All of us wear a mask of some kind that covers varying degrees of our true selves from others.

This mask is the image of ourselves that we present to others. It is our false identity that was developed in response to an unsafe and demanding environment, perhaps out of fear. We have different reasons for using masks. We may want to protect ourselves from getting hurt or rejected by others.

We may want to become what others want us to be in order to be accepted by them. Perhaps, we feel no one would like or love who we truly are, so we hide our true selves; or we might not like ourselves, so we pretend to be like someone else. This creates and keeps us in a state of isolation, and that can be deadly. We have become masters at wearing multiple masks, manipulating, and keeping others from seeing who we really are. But this works against us, because we become lost behind our own masks. They keep us in a state of denial and hold us back from our genuine selves. It prevents us from gaining the much-needed help from others and even keeps us from seeing our own need for help.

When the mask becomes our perceived reality, then our troubles begin. We confuse the mask with the person, and if we are great actors, so do those around us. Gradually the mask becomes a trap, and we become the mask and not our true selves—free to be imperfect, vulnerable, or proceed with bravery. When we have become so comfortable hiding behind masks, we often don't realize we are wearing a mask at all.

How do we find balance? Some ways are to explore your inner depths and learn more about the true and false you. Begin to distinguish between the two and discover all you can about yourself—not only your good qualities, but also your unhealthy qualities. You must explore your dark side and find healthy ways to express what you find in the shadows. Work to appreciate and accept your qualities, style, strengths, and weaknesses. This is the imperfect but real road to recovering from anything and rising from it and through it. Learning about what masks we wear can be the door that opens in finding the balance and the assertiveness to our real selves.

Allowing your true self to come forth is scary because to do so risks rejection. Take small steps such as writing your feelings down, talking about them with a trusted friend, and sharing with another safe and secure person the answers to your questions. To recognize, accept, and share (R.A.S.) is a small but mighty step in letting the masks come off and in your risk-taking to feel safer. Some may have so much pain and anger from past rejections or betrayals that they may need professional help. Do not let your mask suffocate who you really are, but rather use it to create safety and security so the rest of the world can enjoy the real you.

What kind of masks do we wear? Our masks can change to meet the demands of our surroundings. Those who wear masks on Halloween, for example, are in disguise only on occasion. Pretending to be something or somebody else on Halloween we know is a game of pretend. If we are aware of our masks, then we know we are not what we pretend to be. But many are not aware of the mask they present to others.

It takes a lot of energy to keep the masks on. The gift of recognizing our masks and taking them off is congruence—having our feelings, words, and actions align in harmony.

We can harbor many secrets. In doing so, we hold hostage and betray our healthy sense of self, talents, abilities, hopes, and dreams. For years, many of us have covered low self-esteem by hiding behind phony images we hoped would fool people. Unfortunately, we fooled ourselves more than anyone. Although we often appeared attractive and confident on the outside, we were really hiding a shaky, insecure person on the inside. It takes courage, honesty, and vulnerability to begin to remove the masks.

Here are some of the masks or disguises we wear:

- **The Mask of the Victim:** "As long as I wail about how helpless I am, well, I just don't have to take personal responsibility." This is not to confuse the fact that many of us have been victimized. Therapy and support can help us through this. The mask is worn destructively when we continuously use it and do not seek help to relieve our struggle or suffering.
- **The Mask of Silence:** "If I speak out, I might cause someone pain or discomfort, might rock the boat, might actually have to take a stand or take some action."
- **The Mask of Blindness:** "I'll refuse to see the things not working in my life and put a smile on that says everything is okay. Surely then, the problems will go away."
- **The Mask of Coolness, Aloofness:** "If I actually revealed my feelings, I would have to own them, experience them. Way too scary!"
- **The Mask of the Social Chameleon:** "I want everyone to love me and approve of me; therefore, I shall be everything to everyone."
- **The Masks of Busyness:** "If I slow down, I just might have to face the fact that my life is out of whack." (chaos)
- **The Mask of Stuckness:** "I really want to change but 'I can't,' 'I've tried' and as hard as this place is, in truth, as long as I stay here, I am in my familiar comfort zone."
- **The Mask of Morality and Judgment:** "It is easier for me to see everyone else's flaws and shortcomings than to look at my own."

- **The Mask of the Daydreamer:** "I am a visionary, a highly creative individual who thrives on planning extraordinary things for the future...but please don't expect me to apply any of it to the here and now."
- **The Mask of Self-Sacrifice:** "The way I earn my worth in the world is determined by how good I am at shelving my own needs to provide for the needs of others."
- **The Mask of Confidence:** "I have everything under control, I am powerful, and nothing can throw me off my game."
- **The Mask of Having Everything Together:** "Everything I do turns to gold. I do the job right, and nobody else can see me or my actions being a problem."

I encourage you to take time and write down the answers to these questions. It can be an intriguing and insightful exercise. Why we behave the way we do is mostly taught and believed, but it doesn't necessarily mean it came from a great role model or teacher.

1. What masks do you wear? (List them.)
2. What do each do for you? What's the payoff? (List for each.)
3. What are you hiding behind those masks? (List for each.)
4. Where did they come from/when do you remember putting them on?
5. What would life look like without those masks?
6. What are some steps you can take toward a life without masks?

Masks, thoughts, feelings, anxiety, destructive behaviours, perfectionism, people- pleasing, addictions, trauma suffering, and generally most human suffering can be helped. Being a human is not easy, but turning away from ourselves by avoiding, stuffing, or running away from the discomfort is one of the major causes of all human suffering.

You will discover that once you answer the mask questions, your fears, insecurities, anxiety, and stress begin to be understood. What you will also become aware of was that most of your fears were constructed by someone else. Here we return to our origin, the root cause and likely the place to begin rising from. Our masks are misunderstandings. Our life as adults is now our own, not the misguided fears or personas of someone you learned them from. I'll suggest to clients that they look at the person or people that taught them those masks and see if indeed they are still wearing them and how their life is turning out. That question and your answers can provide insightful observations into the past that will encourage you to rise from it.

A client discovered the mask of coolness was learned by her in a long-term relationship. The person she was with for five years never talked about their feelings. Her partner stopped listening to her when she expressed she was angry, sad, or worried. He would say to her, "Get over it. It's not that bad," or "Have a drink." It became a form of inner and outer rejection over years of not having her feelings being listened to. She numbed, withdrew, and lost her self-worth. Depression and anxiety set in.

Feelings and masks are indicators our emotions and behaviours talk to us. Behaviours and feelings seek loving attention and respect. At some point in our lives these feelings and masks kept us safe, but as adults, we may be in a relationship with self or others that has us emotionally underdeveloped and unexpressed. A forty-year-old can certainly act emotionally like a fifteen-year-old. And a twenty-five-year-old may never grow up emotionally and be living an empty, lonely life as a sixty-year-old. Our age has nothing to do with emotional maturity or fluency. It certainly can be learned at any age!

Traumatic experiences are defined more clearly by understanding it is not what happened to us but the reality that we experienced it alone. Therefore, the mask and psychological guards stand ready! Likely no one was there to help us process, understand, regulate, or calm us. So traumatic events or feelings are lying dormant which can leave our self-respect waiting for the approval, acceptance, belonging, or attachment from something or someone.

Addiction (behavioural or substance), for example, is an accelerated and immediate way to create a form of attachment. It is a temporary relief from the feelings of the lost child who was not given the reverence he or she deserved...what is happening within us versus what has happened to us. On the opposite spectrum are the habitual behaviours of avoidance (feelings, communication, and honesty) that often can help us feel safer from the attachments that were never formed. Often our trauma is overly cautious of commitment, so it avoids, seeks safety, overcommits, and stays in unhealthy relationships or unfulfilling ones. Even when a person decides the relationship is not working or healthy, they usually stay far too long to maintain some form of attachment and false hope and also avoid fear of failure.

Naturally on the contrary to reverence for the people I see, their self-respect or self-esteem is based on the destructive self-fulfilling prophecy of, "I am not good enough." I am not good enough says, "I do not belong anywhere, and I am not worthy."

I introduce reverence to my clients by looking at their pasts and how they made it through. Because our body and thoughts can be in high gear, we start by slowing down our current connection to our feelings. We begin with months of long practice answering the question, "How are you feeling today?" This question addresses the emotions instead of just asking, "How are you today?"

At the beginning of sessions, I introduce the client to a feelings wheel. Amongst thousands of emotions available to us, many of us experience only a few on a daily basis that we can identify. We get stuck in describing feelings as "fine, good, and okay." Many people genuinely and intellectually know they have feelings but identifying them takes intentional work and awareness.

Emotions can help us feel relieved or calmer when they are moved. This teaches the brain, body, and emotional responses that the danger is not real anymore, or the perception of danger can be moved through by expressing it. By identifying and moving our emotions, they can and will change.

We do not have to relive traumas, but by gently observing ourselves and how we went through our most difficult times, we can discover some incredible qualities that were left unknown.

My client, Megan, initially came to see me for help with trauma-related alcoholism. Megan had attended my outpatient treatment program, and she had also attended our one-year aftercare program. Now almost sixteen months into a healthy recovery from the effects of alcoholism, Megan found herself impulsively yelling at her partner for not doing or saying the things she wanted. It was only a few days prior that Megan's landlord had unexpectedly given her notice to move. The house she and her partner and daughter currently lived in was now going to be given to the landlord's daughter. Over the same week or so, Megan also shared that at the playground she would become annoyed or angry at anyone who came close to her daughter. I invited Megan to look back over the years and her behaviours with her five-year-old daughter whom she protected excessively. Prior to recovery or aftercare, Megan completely accepted this was normal behaviour. Megan shared throughout her early recovery period that at any time if anyone asked if they could spend time with her during the weekends—which in her mind was designated family time, she would become very defensive and angry at the person requesting her time.

Megan was phenomenal at her commitment to one-on-one counseling. This is an extraordinarily effective and progressive therapy for people working through trauma—and with, of course, the right counselor. It had taken her years to learn these defensive behaviours, and now that the alcoholism was cared for, Megan could begin to discover the real reasons she self-medicated with alcohol for over fifteen years. Megan grew up in a home where there was very little emotional stabilization. Understandably, she falsely protected her fears and projected those onto her daughter even though they were not based on her current situation. In fact, she provided the complete opposite for her daughter: a home where safety, comfort, security, and connection thrived.

Prior to recovery, treatment, and aftercare work, Megan had no idea why she behaved in these ways. In fact, she was extremely hard on herself, saying it was her personality or she needed to change her thinking and beat herself up psychologically for being a bad person. Her self-esteem was in the basement, looping endlessly on the "I'm not good enough" cycle. This is a primary reason people turn to self-medication or mood-altering behaviours. I often say to our outpatient participants, "There is no difference between a deck of cards, alcohol, staying in a bad relationship, using marijuana, or excessive video gaming, vaping, or cell phones. These are all ways we avoid discomfort emotionally, but all it really does is keep that self-worth in bankruptcy.

The self-medication/addictive or obsessive behaviours protected everything we learned or could not resolve, and for all the right reasons, as children, adolescents, or young adults. But now that we are in adulthood, it takes time to unlearn and understand where and why these behaviours are there. Just because we are a certain biological age does not mean we are emotionally, mentally, or intellectually the same age.

Our childhood molded us, restricted, and designed involuntary or unconscious behaviours. Behaviours are constructive or destructive—this is mostly dependent on stages of development as we grow.

Slowing down and connecting to our emotions is transformative work. As we begin to respect our feelings, our pasts, and resilience, we can slow our lives or our thinking down when we need to and simply bring the pace down a few gears. Reverence can enter into our awareness.

Once we begin taking off our masks, expressing our feelings, and practicing our focus on what we can control, we can come out of the cave knowing congruently that the "bear" is not going to eat or attack us.

The payoff for our hard work and rigorous self-honesty is a reverence for ourselves and "what will you do with this wild and precious life," as Mary Oliver writes. Rising into our brave and beautiful lives requires a reverence for remembering our roots. Many people fall short, rising too fast without the roots of their heart and/or their emotional pain solidified into the center of the earth. Here the storms can come, the waves can wash over us, but we hold our ground and stand solid as a cedar. We rise, rooted.

Where we came from provides wisdom as to what worked and what we can do differently. It leads us to develop a resilient courage that can accept discomfort and embrace an exhilaration for life that we have likely never known.

Root, Rise, and Roar with a Red-Tailed Hawk

Over the course of twenty years, I have walked and lived on what feels like every blade of grass which grows on a friend's 700-acre ranch. Tucked into the base of the Monashee Mountain range in the North Okanagan regional district in British Columbia, is a place of wonder which continues to be my wise and waking teacher. Pristine Canadian natural beauty, home to naturally roaming horses, coyotes, whitetail deer, elk clans, wolves, and winter tracks of the illusive cougars have provided me with footsteps, guidance, and wisdom. These walks remind me of my rooted footsteps. Walking beside life, we are consistently preparing to rise, and when needed, we will have our wings and fly.

Reverence Learned from Red-Tailed Hawk

—Written and Lived by Dawn King ©

Stay alert.

A red-tailed hawk dive bombed me while I was sitting in the meadow observing a scurry of Columbia ground squirrels. Of course, it did—I was amongst one of the hawk's daily lunch specials. It occurred to me I was not staying alert to my environment. Beyond the basic needs of food, water, and shelter, human comfort zones are overrated. Stuck in the survival brain of safety, I have historically stayed in relationships or locations far too long only to suffer slowly and needlessly, waiting and wondering when *they* will change. To the hawk, I was a steak marinating in comfort. He was probably thinking, "Wake up, get up, or change. If not, I will eat you!"

Grief never dies.

One of the horse herd leaders died naturally today. The hawk sat high in the fir tree observing as if graciously partaking in a ceremony of remembrance. As the horses circled around sniffing the dead gelding, the hawk seemed to be a witness to sorrow. Loss is a side effect of living, and grief can feel like a heart-breaking emotion—a reminder that by risking and attaching to love, the feelings and the healing of the wound is a lifelong process. Grief never leaves, because love is never lost.

Patiently perch and plan.

Constantly, the hawk has taught me patience. Early in life, I would have preferred patience to come quickly. The hawk teaches patience by perching and planning by the river. Sitting for thirty minutes, the rhythm of the land revealed itself. I see a vole out of its hole, a rabbit to the right, a snake to the left, and now a magpie dropping by for a drink. The opportunities were abundant, but the hawk waits. A keen observer, the hawk was not hunting at all but gathering the necessary maps and routes for future nourishment. Patient observation is powerful planning.

Slow down to soar.

Red-tails enjoy soaring in the thermals. I think of thermals as the invisible rivers of the winds of wonder. I have no idea where the winds are taking me, yet I'd like to think I know where I am going. Playing in the thermals and observing the hawk soaring in the solar heat helps me slow down enough to restore faith in the unpredictable winds of change. With a calm certainty, I know the winds are steering me. Faith in the winds of change playfully soar a soul into the unknown.

Be selective when searching for a mate.

Love is not at first flight. I married twice, and discovered within myself years later that I married to feel a distorted and underdeveloped sense of safety. My father married seven times, and my mother married three. Based on generational trauma and healing, I did learn a little sooner. I have gotten to know the hawk's relationship cycle over fifteen years. Red-tailed hawks search for a partner for three to four years and then mate for life. The hawk couple returns to the same nest each year, which they both make together, and they raise their young together, equally sharing in all responsibilities. Being selective pays off and can provide stability. I believe in love birds.

Get comfortable with solitude.

Hawks are cooperative and comfortable being alone. Red-Tails have been silent companions on my walking and hiking trails, helping me to be more comfortable learning to be a solitary human. Loneliness is real but hawks have taught me that all creatures great and small can help me to feel at home and settle into silence.

Be beautiful and brave.

Hawks are beautifully stunning with beefy bodies and rusty red tails. Typically feisty, fast, and protective, red-tailed hawks are a yummy chocolate brown from head to toe. They are ruthless all-purpose hunters, with beaks that rip and tear and muscular legs attached to talons that make my knife and fork seem pathetic in efficiency. Hawks scream like hell. The most epic battle calls and cries I have ever heard from the winged ones has been between a hungry red-tailed hawk and a ruthlessly protective raven. We all have a voice—use it bravely and beautifully.

*   *   *

Trauma, anxiety, and mood disorders can and do heal. The red-tailed hawk offers simple wisdom by being in nature and modeling her unpretentious and profound ways of self and co-regulating. When we slow down, we can heal, and the reverence we feel or begin to see and find within ourselves and in our lives will return, restore, and renew.

*   *   *

> *"I feel that adolescence has served its purpose when a person arrives at adulthood with a strong sense of self-esteem, the ability to relate intimately, to communicate congruently, to take responsibility, and to take risks. The end of adolescence is the beginning of adulthood. What hasn't been finished then will have to be finished later."*
>
> —*Virginia Satir*

CHAPTER EIGHT

Growing into Congruency-Trigger less through healed trauma

Congruency is both a promise and a practice; a lifelong journey toward aligning with our authentic selves.

When we begin to feel calm instead of chaos inside, we feel safe internally and externally. Growing into Congruency invites an ease into emotional and physical regulation, reorganizing chaos into calm and turning drama into dreams. I think it is absolutely possible for us to heal our pasts. The journey is not easy, but it actually is quite simple. In order to root and recover from the past physiologically, emotionally, and physically, we have to grow up all the parts that taught us to limit our capabilities. We then rise into our real selves and roar our authenticity unapologetically. We now take up the space that felt unloved, unworthy, and unregulated with solid self-worth, emotional fluency, and a healthy relationship from the inside out. Congruency soothes and regulates the effects of anxiety, addiction, or trauma and all those little or large lies about who we are or who we are capable of becoming. It can literally be the antidote for that pain.

Congruency brings our thoughts, actions, feelings, values, and all the parts of our whole selves into an integrated alignment without having to really think of them. We slow down, and our words, thoughts, and actions leave us feeling honest and confident. There are no masks, no disguises, and no adjusting to different people or environments so we can feel as if we belong or are accepted. We feel the connection and acceptance of ourselves inside first. There is a gentle, connected, and graceful confidence. There is a pause to ground our whole selves, and it is attached to calming and listening to our inner voices before we say anything, think anything, or consider an action.

Congruency is a direct connection to self. When I deliver the parts of me out into the world to an individual or a group, it simply means I express what seems to be the most honest parts of me. Free of people-pleasing and perfectionism, I accept all of me.

One of the benefits of Congruency is a sense of peace and calm. There is a trust within us that is practiced and earned. There is power in the pause. Before we speak, do, or respond/react, we learn to pause long enough to answer the questions: How do I feel? Am I afraid, brave, certain, courageous, sad, uncertain…? One of the most effective ways to start practicing the connection to our Congruency is to put both our hands on our hearts, close our eyes, and breathe into our hearts and bodies.

Slowing and connecting to our inner voice and wisdom, congruence asks, "Is this feeling in the past or the future? Where did this come from? Who did I learn it from?" A powerful question I ask is, "Was the person I learned it from a decent role model for what I am afraid of?" Usually, they were not.

Many of us are not aware that we repeat the beliefs and actions of those most influential in our lives, past or present. When we pause, we can ask these questions and see if, in fact, that person is congruent. Are they actually doing what they say and saying what they do? Are they receiving the outcomes they want? Are they walking the talk?

Listening to advice from people who are not congruent won't help us grow. When we go to invest our money or invest in our health or grow relationships or businesses, are we taking advice from someone who is actually succeeding? Congruency asks questions to determine the next step. Those we learned fear from, let fear stop them. Look at their lives currently, and you will see that their actions speak louder than their words. If instead they used their fears to grow and become free from them, that's a decent role model. When it comes to matters that are not life-threatening, our limitations are, for the most part, learned but often not accurate. Our defences lie there, and fear prevents congruence.

Congruency asks us to remain connected and attached to ourselves. Happy or fearful is neither right nor wrong. We become aware that feelings are accessible, which creates an emotional fluency leading us to know where we are on our own path. Feelings are road maps to share with us, and through feelings, we strengthen the trust in knowing which direction to go, when to get off the highway, or when to try something new and head down a road filled with unknowns.

The benefits of congruency include a consistent sense of self-acceptance of where we are and a connection internally first. As in all learning, we have to think about it and practice it until it becomes consciously competent. There is an inner confidence which no longer lives in hiding, avoiding, or holding back. There is an increase toward full acceptance of who we are even if we become aware we messed up or made a mistake. We accept and are accountable for those parts as well. The "I am never enough" we have been housing and hosting for most of our lives lessens in control and gracefully dissipates. Truly, there is no risk in returning love to yourself.

Naturally, we are not congruent all the time, but progress over perfection is realistic and rewarding as we change, learn, grow, and adjust along with other people and new environments. When storms come into our lives, we may not always meet them with congruency, but with each storm, we learn more about who we are and what we need or how to ask for it. We are growing, and with each growth spurt we find or become more of who we are, learning who we want to be, while acknowledging the amazing parts of us already here.

When we are emotionally congruent, we don't have the emotional highs and lows. We don't have to sit in depressed isolation or overachieve looking for external validation. We don't have to use mood-altering substances or behaviours to numb or avoid, because we start to become comfortable with the discomfort and seek care for the wounds or what we did not have. We also don't have to attempt to stay in a fake positivity telling ourselves that all we have to do is change our mindset. Shit happens, things do not work out, and most of us need to take care of and understand the needs of our little kid in an adult body. Congruency takes action on befriending our feelings. We allow ourselves to be real, to feel, and to use feelings as a compass toward what we want and what we don't want.

The more we become congruent, the more we become ourselves. The more we understand ourselves, the more we become congruent. We find a sense of belonging inside and out. Healthy relationships form within us, and we begin to experience them externally as well. The more we become ourselves, there is an ease. Triggers settle that used to have us fleeing or fighting, and we begin to regulate the dopamine in healthy ways and balance our hedonic set points. We become willing to accept all the parts of our whole. There is a settledness in Congruency.

The self-discipline we receive from striving to be congruent is a quality worthy of commitment and self-exploration. Rigorous honesty is required for congruency. We consistently explore and accept our personal inventory, emotions, thoughts, and actions.

We can always retract what we do or say—we make mistakes, we learn, we can do better, we lose our way. This is also Congruency in action. Without Congruency, we simply wonder or avoid who we are and where we are going and let life happen to us. Indecisiveness and procrastination are strategies for avoiding being congruent. If we just say we are a procrastinator, those parts of us remain unveiled, and this is where we end: stuck in reverse.

I am not a believer in "the power of now". It simply has the potential for emotional bypassing. Trauma responses looking to find relief need to move—physically, emotionally, mentally, and spiritually. Too many clients have experienced and share with me their needless suffering and perceptions of failure when they cannot "stay in the moment". Is this concept welcomed by the people seeking more calm and less chaos, or those who seek the meditation or enlightenment mountain? Maybe? A culture that force feeds the cure-all as meditation and mindset are powerful marketing strategies that can fall short. We don't hear about the people it doesn't work for. Hopefully, people whom these strategies do not work for find a great therapist or another path toward truthfulness and congruency based on their individual selves and for themselves. If you are one of these people that continue to seek help that does not work or sustain, you are not alone and you are not doing anything wrong. One size does not fit all.

Trauma likes to love itself, and that is naturally how it is resolved. The fight, flight, freeze, or fawn responses often occur trying to stay in the moment, and it can actually trigger a traumatic response. A woman fearful of snakes (or anything) is told, "Just hold the snake, pet the snake, stay in the moment, and be one with the snake," to overcome her fear. Nope, run away from the snake! Moving and not staying in the moment is the solution. It must allow space for other ways of finding self-acceptance and emotional honesty. We are always moving, and life is unpredictable. We simply take stock of where we are, what we are feeling, and take the risk in actively proceeding in any direction. Congruence is a moment-to-moment, or situation-to- situation, or relationship-to-relationship, way of being. It is rarely stationary or stagnant in any given moment.

What stands in the way of congruency? Not trusting ourselves because we fear failure, success, change, what other people think, or getting hurt. People-pleasing, perfectionism, and not standing up for ourselves hinders the growth of congruence. Congruency asks us to take risks in finding who we are and what we want. Either we risk being our real selves or we risk not being our real selves. Both have payoffs, whether constructive or destructive. We can stay in a relationship (with ourselves or others) miserably, or we can be honest and say it's not working and change it. We can acknowledge that we procrastinate and use this as a protective excuse to not follow through, or we can be bravely honest in self-exploration and be real in seeking to understand what or why we are avoiding. Both take risks, but both have completely different outcomes: stuck or free.

We are control and comfort-based. Most of us were raised to fear change. Therefore, we stay in our caves waiting for the bear or lion to arrive. Most of our fears regarding failure and change can be overcome with a simple practice and question: what can I control? The simple answer is our thoughts, feelings, actions, and behaviours. That's it. This is the practice of congruency.

As soon as we get out there trying to control others—as in advice-giving, worrying how we will make them feel if we are honest about our own needs, disappointed if people don't do what we told them to, or so attached to what other people do or say that we take ourselves far away from our own wants and needs—we are in a cycle of dependency on others. Destructive dependency means whatever others do has a profound and lasting effect on our well being. Yes, people will hurt us. If we carry that hurt or trauma, we have a personal responsibility to be accountable for our pain. If the person we want to control doesn't agree with us, we believe we must be wrong. I'm ok if you're ok, it not going to be an ok outcome. If they don't change, we better try harder, again and again, to change them. If they leave us, we are a bad person. If we didn't get the promotion or job, we might never risk going for what we want again and stay in work or a job we absolutely loathe because it pays the bills.

We may have not caused our pain or trauma, but we have the right as humans to care for ourselves and be cared for. We are responsible for our pain. We are accountable to caring for it and asking for help.

Beyond food, water, and shelter is where we want to bridge the survival brain to emotional or behavioural Congruency. Many of us have food, water, and shelter, but we exist as if we are still in survival mode, and our anxiety may be high. We would all benefit from learning that we can't control what others do, think, or feel. We cannot control whether they choose to interact with us, whether or when they choose to grow and change, and whether or when they choose to recover from addiction, save money, go to college, get a better career or relationship, or endless other scenarios.

While writing this part of the book during the summer in British Columbia, Canada, where I live, our communities had been surrounded for three months by historically high temperatures and wildfires burning in every direction. Bags were packed and ready to go at a moment's notice. Our city experienced steady power outages and no relief from the heat or the fires for months. What could I control? My car had a full tank of gas, the safety of my cat was secured in a kennel, I had water in storage, the food was prepared in the pantry and deep freezer, and I had numerous filled propane tanks to cook food with. Countless times there was no cell service or power, and I continued to type words but was not able to control the battery on my laptop from eventually running out.

We are also not in control over the backlog of feelings and negative beliefs we have packed into our minds and housed in our bodies. Sometimes or frequently the past comes out through our addictions, anxiety, depressions, relationships, or what people refer to as the baggage we carry. We cannot change the past, control our friends, family members, adult children, or other people's children. We really cannot control every result we strive for, nor can we control all life circumstances or events. This can be freeing or terrifying!

We have a false sense of control by rigidly and maybe unknowingly repressing our feelings and thoughts. With this behaviour comfortably in place, we lose ourselves. When we try to control people, it is exhausting. It is a distraction from our own change or self-awareness. When we try to control others' behaviours, their behaviours control us and how we feel. When we try to control what others think of us, we turn into the directors of someone else's life and learning. The cycle repeats. Attempting to control other people's thoughts, feelings, or actions puts us under the control of whatever we are trying to influence or change. This creates disconnection from our own sense of self and distracts us from understanding what is congruent and real for us.

When we try to control situations and circumstances, we set up blocks for ourselves to move forward. When we spend time and energy trying to have power over people, then we have little for ourselves. Examples of thought patterns include: If my partner was better at finance, we would both have more money, or, if my father or mother changed, my trauma and anxiety would go away. We may long to control or carry false hope that others will change. This is where believing we can control others' changing can be emotionally defeating. Our empowerment and confidence build as we care for ourselves and our pain and remember how far we have come.

We impulsively want to control others, especially when they're hurting themselves or us, or when things aren't working out the way we had planned. Letting go of control and finding congruence allows us to accept that it is not our job to fix others, their feelings, thoughts, decisions, growth, and responsibilities. Our adult children may want to change careers. It is up to us to stay in our lanes and let them find life for themselves. Regarding our adult children, wait to be asked. Most of them spend nineteen years (or more) under our guidance and control, and they really are ready and want to fly from the nest. Yes, they will get lost, and yes, they will find their way. Picking up their toys and cleaning their room eventually becomes our children's self-responsible role in life. This is how they find their way—by doing and facing consequences and reaping the benefits of overcoming adversity.

We may want to keep telling our partners to take care of their bookkeeping or actually do their bookkeeping because we feel if we can control their responsibility we will feel better. The cue is to see how resentful or frustrated you feel by actually doing their bookkeeping. Our feelings guide us into the awareness that this is not our job, or we may do it joyfully and willingly. These are two significantly different control situations and feelings. Being congruent asks us to do that for ourselves. We have been groomed and conditioned to listen and repeat another person's false sense of comfort and safety.

It's our parents, our culture, our teachers, every attachment to another person or reference that says to hold back and be afraid to be our real selves. We are in the hypnosis of approval and acceptance from others.

We are responsible for ourselves, and that is plenty. Others are responsible for themselves—whether or not we like, approve, disapprove, or think we know what is best for them and criticize or harshly judge how they are handling them. This wears a relationship down and gives the message that this person is not capable of doing it themselves. Enabling moves us further away from being congruent.

When we become open to seeing how much we get an imaginary and temporary comfort from attempting to control others, we start the congruence process.

What we are in control of is stopping our own pain, facing and dealing with our own fears, saying no, giving ourselves what we need, setting boundaries, and making the choices and decisions we need to make to take care of ourselves. This is growing ourselves up from the past and creating congruent acceptance of all parts of ourselves. We can certainly control changing the parts of us asking for a change if we are willing to slow down and listen.

When we stop controlling and seek congruence, life on life's terms will fall into place. When we take control of learning more about our thoughts, feelings, actions, and behaviours, we find our place in our own world. Belonging there is our first step to belonging anywhere.

With time and practice, we eventually develop reverence, appreciation, and acceptance for the ways people make their own choices and decisions and how situations work out because they may be better and different than what we could have accomplished with our controlling behaviour. We can be free here and so can the other person.

To be emotionally free—to untangle and unshackle the pain and defeats of the past or unlearn anything we don't want—requires honesty, patience, and bravery. Courageous congruence is a pathway which explores our inner selves with fierce curiosity and seeks greater meaning and purpose for all that has occurred. It's as if the good, the bad, and the ugly thread the needle to a finer and more beautiful form of tapestry. We all experience trauma, tragedy, despair, and anxiety. Most of us have also struggled with some form of addiction or obsession. Most of us have stayed in a relationship too long or not put the 500 miles in to making the ones we have better. Most of us have impulsively acted in a way such as spent too much money or temporarily relieved our emotional states by Netflixing and avoided life's discomforts in countless behavioural denials or addictions.

Yet, in real life, we also experience triumphs, and the gifts of life's adversities are determined by their meaning, the messages or feedback those adversities can give us, if we are open to the possibilities, and seek the actions for necessary changes within us. As American author and disability rights advocate Helen Keller reminds us, "Although the world is full of suffering, it is also full of overcoming it."

Congruency is a process of measurement and a check-in as to why we are doing too much or not enough. The myth of more is when we discover that avoidance is a primary strategy, and we do more of almost anything while destructively falling into habitual behaviours that move us away from our real selves. Striving for congruence allows a connection to self, an inventory process for inner inquiry to determine if indeed our decision and actions are in alignment with what is most important to us. The payoff of congruence is that we will naturally create constructive behaviours.

Following through on our word is a simple but profoundly effective example. If we say we are going to do something, in congruence, we do it. Yet how many of us fall back on the "procrastinator" default strategy? As psychologist and psychiatrist Carl Jung states, "You are what you do, not what you say you'll do."

Congruency is a form of harmony between our goals, dreams, quality of relationships, desires, beliefs, values, and mission. A person with congruence does not necessarily live in a state of Zen or in the power of now or engaging in meditation on some mystic mountain—quite the opposite, really. Congruency takes brave action and a vulnerability to discover the light and dark sides of ourselves. Author T.S. Elliot spoke to this when he said, "The darkness declares the glory of light."

This is a process that we can use daily to check in on our congruence and see where we would like to grow it. The following framework may help to integrate the practices and principles of Root, Rise, Roar.

ROOT

Admit or Acknowledge

Admitting our strengths, pain, hopes, dreams, or our feelings can be easier than accepting. We can talk and intellectualize knowing we could or should change, but doing something about it is the continuum of congruence. We must actively move toward acceptance of that change (or not) as it is necessary for our emotional peace. Root, admit, and acknowledge into understanding your feelings and the Emotional Medicines chapter. Root and ground yourself into Mother Nature—the original therapist. Mother, nurture and connect back to yourself, through her. Root your entire sensory system into slowing down your crazy child. Admit and acknowledge you may be going this way too quickly and not arriving or even knowing where you are. Slow down and determine what is not working and admit the cycle and speed needs to slow down so you walk alongside yourself.

RISE

Acceptance
The gap between admitting and accepting can be a life changer. Accepting has activity, and actions will speak and provide more results than our words. By overcoming one defence mechanism or one controlling behaviour at a time, we move our ability to change from admitting to accepting. Here we can actually take the next step, which I call R.A.S.—Recognize, Accept, and Share.

We need to RECOGNIZE our dark side, our imperfections, areas to improve on, shortcomings, ineffective actions, and destructive or stuck emotions or behaviours. Sometimes we forget to recognize the great and whole parts of ourselves. Many people fear success more than failure. What if I were to be the real me? Can I accept failure and success?

Now SHARE those constructive or destructive parts with nother person, a plant, or trail or run or yell it to the skies. Admit it, and announce it, just stop avoiding it. Avoidance causes anxiety—it is the ultimate lack of resolve. The "what if I do/don't" will never be known if you don't admit, acknowledge, and accept the problem or the solution. Denial and delusion live in the same closet. Believing something that is false to be true and not acknowledging what is true creates a relentless cycle of needless misery and suffering. Releasing control over people will bear this fruit—"admitting" where we really are, that we have ineffective and possibly harmful or destructive defences, and/or simply that things may not work out.

Accountability is the ability to accept all aspects of our behaviours—trauma, anxiety, confidence, assertiveness, all of it. We are accountable for traumas and triumphs. We have the choice to recover from hurt and heal.

As difficult as it is necessary, this stage of acceptance is emotionally liberating. Awareness, admission, acceptance, accountability, and sharing it with someone we can trust and are safe with will help us get to a place of healing.

We then may hit the wall or see the stuck cycle returning. The wall feels mighty, but we are mightier! What makes the "wall of stuck"? Three behaviours:

- AVOIDING—thoughts, feelings, communication, or behaviours. As we understand the difference between constructive and destructive behaviors, avoiding leaves our roots rotting in denial. We cannot grow from there. We destructively ROOT and rigidly do not change, causing anxiety, depression, addiction, and other destructive behaviours we have the power to change.

- APPROVAL—from ourselves and others. When we deny, neglect, or abandon ourselves and seek or wait for approval from others, we hold our real selves back, shackled in what other people think and limit our spirits from soaring or roaring. We destructively stay STUCK at the lowest common denominator of those around us or the limitations we regurgitate within ourselves. Others' approval prevents us from RISING from healthy and strong roots.

- APOLOGIZING—holding ourselves back by being overly responsible for everyone and everything. I am Canadian, and I look forward to the day that apologizing is not seen as one of our strengths. Humility and respect, yes, but apologizing is not a strength when we have done nothing wrong. Then when we do make a mistake, apologies have no backbone. Constantly saying, "sorry" lowers self-esteem, and it's annoying because it has no relevance, substance, or authenticity.

How do we jump over, crawl around, pole vault, or break through the wall? Incorporate the use of "The Secret of the Soma" and sing about it! Be Mindful of Movefulness and move your body and release, renew, and restore the feelings and experiences previously shackled. Through Singing and Movelfulness, you can activate those trapped emotions. Stifled or stagnant feelings will release and rise through self-expressions by writing, painting, journaling, dancing, sports, engaging in social activities, and moving our bodies. Be willing to ask for help and share with someone you trust and feel safe with.

ROAR

Action

Congruency releases all control over others. When we live in authenticity, we let go of advice-giving, fixing, enabling, and allowing our imagination and creativity to permeate all that we do or stand for. It is a stage that lasts until we no longer live physically. All other stages of congruency merge into a tapestry of known and unknowns. The real me is more comfortable in taking risks, crossing thresholds, and overcoming adversity, and has become accepting of uncertainty. The greatest cause or fuel of anxiety and addiction is control of that which we cannot control. Those who can settle into the unknown find their true nature. Lost in the forest, how do we find our way home? Through these three stages of Root, Rise, and Roar.

Once we begin to create emotional fluency and identify numerous emotions available to us, we can use our emotional energy to spring into action. All the stages up until now have been preparing us, and now taking action toward change or recovering from anything will return to us the resources and opportunities we may have never imagined. This is a stage that cannot be missed but people can find the most difficult. Think of it this way. No matter what we do or don't do, we are taking a risk. Higher standards, healthier relationships with self or others, finding Courage through Communication, or changing behaviours is risky. Either we risk more of the same struggle and suffering, or we risk creating a change. Both risks will have discomfort, but the pathways will be substantially different in their outcomes, and nothing is guaranteed. This is the primary reason people do not take action: discomfort in the unknown.

Deep in the Reverence or respect earned for ourselves and this gift of life we can settle, knowing that life is always asking us to build a bridge between comfort and the courage to grow. We can borrow the past and acknowledge the perseverance it has taken to have come so far. How has your adversity prepared you each step of the way to, not only keep going, but also find the help that was waiting for you? When we take off the masks, we find ourselves rising and roaring into our real and authentic selves. We roar our anger, our pain. We resolve our pasts; we remove the masks and reclaim and renew who was underneath it all. From a wounded child to a wise woman or man, we ROAR our Reverence for all this life gives us, including our pain and pleasure. The willingness to actively proceed in Courageous Communication with self and others, while journeying long and far enough that we return to ourselves, is essential.

It is likely you will go back and forth and up and down throughout or between these three stages of Root, Rise, and Roar. They are here as a guide and a stocktaking tool when you are moving through the Root, Rise, and Roar process.

In Congruency, our thoughts, feelings, actions, and words align to a sense of real self. Often the most congruent people continue the path of always learning, understanding authentic goals, and discovering even more layers of who they are and unveiling another part of what is within them. A sense or state of congruence allows a steady intravenous injection into our veins of peace, confidence, and passion.

Congruency is earned, and it is hard work. The payoff is an authentic, peaceful, passionate, trauma- and anxiety-free life, all tucked in with warm hearts and healthy relationships from the inside out. Be gentle with yourself, change and integration is an individual journey, a wholehearted process and a commitment to integrate. The rewards will be worth the wait and the work you put in. You deserve to care for yourself and make your recovery, your brave and beautiful life your number one priority. Congruence releases all control over others.

Once we begin to create emotional fluency and identify emotions available to us, we can use our energy to spring into action. All the stages up until now have been preparing us, and now taking action toward change or recovering from anything will return to us the resources and opportunities we may have never imagined. This is a stage that cannot be missed, but people can find the most difficult. Think of it this way. No matter what I do or don't do, I am taking a risk. Higher standards, healthier relationships with self or others, or changing behaviours is risky. Either I risk more of the same struggle and suffering, or I risk creating a change. Both risks will have discomfort, but the pathways will be substantially different in their outcomes, and nothing is guaranteed. This is the primary reason people do not take action…discomfort in the unknown.

Authenticity—when we live in authenticity, we let go of advice-giving, fixing, enabling, and allowing our imagination and creativity to permeate all we do or stand for. It is a stage that lasts until we no longer live physically. All other stages of Congruence merge into a tapestry of known and unknowns. The real me is more comfortable in taking risks, crossing thresholds, and overcoming adversity, and has become accepting of uncertainty. The greatest cause or fuel of anxiety and addiction is control of that which we cannot control. Those who can settle into the unknown find their true nature.

Lost in the forest, how do we find our way home? Through the previous six stages.

It is likely you will go back and forth and up and down amongst these stages. These are here as a guide and a stocktaking tool when you are moving through the root, rise, roar process.

In Congruence, our thoughts, feelings, actions, and words align to a sense of real self. Often the most congruent of people continue the path of always learning, achieving goals, and discovering even more layers of who they are and unveiling another part of what is within them. A sense or state of congruence allows a steady, intravenous injection into our veins of peace, confidence, and passion. People displaying congruence simply do the right thing. It is obvious as other people tend to want to be around them or feel an inspirational gravitation toward their presence, mission, or vision.

Congruence is earned, and it is hard work. The payoff is an authentic, peaceful, passionate, trauma- and anxiety-free life, all tucked in with warm hearts and healthy relationships from the inside out.

Wendy
R3 Program Graduate

Over the past several months I have embarked on a journey of self growth and enlightenment which has landed me here today to rewrite this letter. Words cannot express how incredibly rewarding this journey has been. Although at times overwhelmingly tough, I am deeply grateful for the fortitude, strength, and courage I have to become the best me I can be. At the beginning of the R3 program, I was asked to write a story about how I perceived my life to be. I titled it "The Drift". After rereading this story, what is profoundly apparent to me today is that for many years I just drifted through life. Waiting for the winds to blow me in a direction that would fulfill me, seldom grabbing the rudder and steering my own course. When something did stir in me to take control, I was so overwhelmed with fear and uncertainty, I would let go and continue to drift. I was miserable. But, after setting on this journey of well being, I learned so much about myself and the fears that have kept me locked in destructive cyclical patterns. I stepped away from blame and self pity and became accountable for the direction I wanted my life to go. I have chosen to adjust the sails and head towards the vast sea of contentment, joy, and fulfilment. I am now a woman of great worth, and although I have more work to truly believe this, I stand tall with my head held high, shoulders back and proclaim I AM woman, hear me roar! I AM trustworthy, strong, determined, courageous, lovable, compassionate, confident, I beautiful and capable, and as I enter elder-hood I will now refer to myself as "Polished Silver", and with continued maintenance and care I will remain "fucking radiant!" Throughout this journey I have learned to improve on my self respect, love, and value, set boundaries and adhere to them (though this area still needs practice), and how to move out of adolescence into adulthood. I've learned the value in reaching out for support and being my most authentic self. I have taken responsibility for the broken relationships in my life, by being accountable and staying focused on what I can do to make things better. I no longer feel resentment, anger, envy and blame. Instead, I have found an acceptance I have never known before, and in turn I now know what unconditional love actually "feels" like. For that I am sincerely grateful.

I have found that learning to walk with and through my fear, while in the beginning rather nerve racking, it is incredibly empowering. I'm finding it easier to embrace fear today. By holding fear by the hand and letting go of self, I crossed over to acceptance and uncertainty. Family is one of the things I value deeply, so taking steps to ensure its health.

Wellbeing is another personal value of mine. Nature, growth, laughter and health are also things I have been working on. I have implemented more laughter and lightness into my days, by choosing less serious topics, and being more playful. I've made snowmen and snow angels, performed in a silly skit, laughed uncontrollably, and howled at the moon. But most importantly I can laugh at myself at a time when I would have felt inferior for not being perfect. I am so proud of my commitment to get back into strength/cardio building, and a better diet, as I can see the positive results because of it. I am stronger on the ski hill, (yee-haw!) My back doesn't hurt any more, my clothes fit better, and I feel better mentally and physically. (whoo-hoo!) I have sustained my connection with the WOW (women of worth) I've had the pleasure to walk alongside over the past six months, as we had agreed to meet once a month. I will continue to support them however they need and reach out for their support when I need an ear, shoulder, or a good swift kick. Especially when I say to myself; "I can do this alone". I am so very proud of myself, and I am beaming with confidence and self love! I am proud of the courage, commitment, and hard work I've shown to get where I am today. My growth is undeniable, and glorious. times get tough, as they will, I will utilize the toolbox I now acquire that is bulging with new effective tools to help me stay on track and see me through. And when that's not enough, I will reach out for support and connection from my balcony friends. The famous WOW! The devotion I have towards my personal growth is staunch and I have pledged to continue this journey of self betterment, by taking

time to reflect, regroup, and possibly reboot. Until then I vow to use the tools I have been given, stay accountable and keep adjusting the sails.

Forever Yours,
Wendy

Tanya
R3 Program Graduate

Dear Tanya,

I am so proud of the strong, independent woman you have become. I can see how you have learned to set boundaries for your own wellbeing. Setting boundaries has given you more freedom in your life, helped support your needs and has created more connection in a healthy way. Where there used to be a flare-up of destructive behaviours such as chaos, drama, story telling, and blame, you have found the courage to decipher what you need from a situation, what your responsibilities are, and how to care for yourself through it. Not always perfect, mind you, but perfectionism isn't what we're striving for anymore.

I can see how you are taking responsibility and accountability for your own part in the chaos making and busyness patterns. I am proud of your awareness and dedication to change these behavioural patterns and can see how you can see your contribution to these destructive ways. I am proud of your choices to make these changes, and the courage it takes to do so. These choices are leading you towards more connection, time, and greater contribution in the relationships you wish to foster - like with your sons, close friends and family members. I can see that you have become more comfortable with living at a slower pace, and factor this in when planning your day. You are listening more closely to the messages and signals around your own personal boundaries. Boundaries related to chaos, over-busyness and activities that consume time and energy. Learning to calibrate before the pendulum swings to any extreme is a milestone worth celebrating!

Getting into nature, connecting and re-establishing your roots has been an incredible tool in nurturing yourself and in asking for wisdom to come through. Your ability to trust yourself is growing everyday. I can see your strength and confidence behind your decisions as you lean into trusting the voice and guidance from within. I am so proud of the amazing accomplishments of overcoming feelings of terror and destructive fear that used to consume and take over the day. When I look at how transformative that is, I get so excited that my heart could burst out of my chest. To be able to manage your emotions like that, is so commendable and really worth celebrating - well done!

You have been shown the way: awoken to the power that lies within the emotions that already reside within you. The R3 Program has taught you so many new ways including calling upon the emotions that lift you up, fire you up in the most supportive and powerful way possible. Whether it is to stand up for yourself, build your confidence or break a destructive pattern/emotion, leaning into strength and empowering emotions has literally transformed you into a strong, courageous, empowered woman. Not that you weren't before, but now you have the ability to tap into these emotions and stand tall, proud and seen. No need to cower away, feel small or needy again. You are not helpless. You are strong, courageous, decisive and wise. A fine woman of exceptional worth.

I am proud of you for finding your strength and ultimately growing your worth along the way. You are beautiful and loved.

Supported and guided by love and God, who is everywhere and everything. Your relationship to self is your root, and nurturing that and tending to the fire within will keep you on this path of awesomeness. Celebrate your worth, standing tall and proud - Heart and chest to the sky, arms and face up to the sun. Feet firmly planted on the ground. Go forth strong warrior, celebrating your courageous journey of R3 and forever taking it forwards with you.

Love,
Tanya

EPILOGUE

SACRED SEAT
A home where you are safe and warm

Dear Traveler,
A healing or renewing journey can take time to acknowledge and accept. It takes honesty, openness, willingness, courage, and vulnerability to slow down, yet still proceed. Growing ourselves up is the necessary threshold to a brave and beautiful life. It is simple and with struggle and not meant to be easy. We must ask for help. If help is not in front of you or around you, try again, and again, and again. Accepting the past and the present, our power, pleasure, and our pain will move us into a congruent and authentic existence, free of traumatic responses, self-harming, or destructive coping mechanisms. Accepting and growing into ourselves compassionately for all we are and where we have come from is the medicine for a trauma- and trigger-free inner world.

As we recover and heal, we transform fear into faith, addictions into connection, anxiety and chaos into calm, isolation into compassionate care and community, failures into feedback for growth, and feelings into our powerful ally. It is then and there, trauma is transformed.

Healing trauma and integrating changes or perceptions within our mind, body, and emotions will untangle our spirits. Take all the time you need.

For those of us who have journeyed and healed—we are here waiting for you. Welcome to a home where you are safe and warm. We will keep the fire lit and save your heart and soul a sacred seat.

Prayer for Our People

May we listen with grace to the voice of the wounded child and to the wise elder within.

May we build a bridge of love so that our crossings can heal.

May we liberate and transform our pain into a brave and beautiful life.

May we speak of kindness and compassion for our wounds.

May our words and actions be our soothing salves and healers.

May we patiently hold each other in the darkness until the moonlight, the stars, or the dawn returns.

May our hearts then be freed to begin again.

May we restore, reclaim, and remember our joy, our faith.

We hear your voice.

We are here holding your pain until you return to the safe chambers of your heart.

—*Written and Lived by Dawn King* ©

R3 EXPERIENCE
ROOT. RISE. ROAR.

Invitation

For decades, it has been a blessing to help thousands of people look through new eyes, see new possibilities, expand, elevate, explore, and transform their behaviours and lives. It would be a delight to meet you. I hope to see you on the path.

5 Years of therapy in 6 Months

R3: ROOT. RISE. ROAR. Offers cutting edge and ancient wisdom through individual and group programs. These offerings integrate the best of my therapy practices with my extensive mentoring strategies. The R3 program is a powerful and transformative combination of mentoring, behavioral therapy and group sessions with a courageous and compassionate community. You will hear the whispers or the hollers of your heart.

- Discover fear is your friend and your compass.
- Give your doubts and excuses the heave-ho.
- Build an extraordinary relationship with self and others.

- Become a cycle breaker and wayshower
- Eliminate people-pleasing and perfectionism.
- Stomp on self-doubt, worry, and anxiety.
- Create emotional fluency and courageous congruence.
- Transform trauma into self-worth and assertiveness.

Ecotherapy and Ecopsychology Experiences

Our programs also include in-person ecopsychology and ecotherapy experiences. We will meet in person as a group in some of the most wild and magical places where we share deep harmony and connection to ourselves, others alongside the mystery and magic of the lands. Rejuvenate and integrate your true nature.

I have written this book from my heart. I am certain your healing, and your growth will be an epic journey into the known and unknown. If you have fear, doubt, or excuses—great! Jump in with faith and courage! May you ignite your fear into faith, observe your failures with growth, and listen to your feelings as your most powerful guide.

May you transform trauma into your brave and beautiful life.

Dawn
www.DawnKing.com

BIBLIOGRAPHY

A., Van der Kolk Bessel. *The Body Keeps the Score: Mind, Brain and Body in the Transformation of Trauma.* UK: Penguin Books, 2015.

"All about Emotional Tears." American Academy of Ophthalmology, April 12, 2017. https://www.aao.org/eye-health/tips-prevention/all-about-emotional-tears.

Blackie, Sharon. *If Women Rose Rooted a Life-Changing Journey to Authenticity and Belonging.* Tewkesbury, Gloucestershire: September Publishing, 2019.

Blackie, Sharon. *If Women Rose Rooted: A Life-Changing Journey to Authenticity and Belonging.* Tewkesbury, Gloucestershire: September Publishing, 2019.

"The Body Shop Global Self-Love Index Market: US." The Body Shop, March 2022. https://thebodyshop.a.bigcontent.io/v1/static/US-Self-Love-Country-Report.

David, Susan. Susan David: The gift and power of emotional courage | TED Talk. TED, 2017. https://www.ted.com/talks/susan_david_the_gift_and_power_of_emotional_courage.

Lindahl, Jared R., Nathan E. Fisher, David J. Cooper, Rochelle K. Rosen, and Willoughby B. Britton. "The Varieties of Contemplative Experience: A Mixed-Methods Study of Meditation-Related Challenges in Western Buddhists." *PLOS ONE* 12, no. 5 (2017). https://doi.org/10.1371/journal.pone.0176239.

"Pain, Explained." Psychology Today. Sussex Publishers. Accessed June 23, 2022. https://www.psychologytoday.com/us/blog/pain-explained.

Pritchard, Alison, Miles Richardson, David Sheffield, and Kirsten McEwan. "The Relationship between Nature Connectedness and Eudaimonic Well-Being: A Meta-Analysis." *Journal of Happiness Studies* 21, no. 3 (2019): 1145–67. https://doi.org/10.1007/s10902-019-00118-6.

Saarman, Emily, and Emily Saarman. "Feeling the Beat: Symposium Explores the Therapeutic Effects of Rhythmic Music." Stanford University, May 31, 2006. https://news.stanford.edu/news/2006/may31/brainwave-053106.html.

Stromberg, Joseph. "The Microscopic Structures of Dried Human Tears." Smithsonian.com. Smithsonian Institution, November 19, 2013. https://www.smithsonianmag.com/science-nature/the-microscopic-structures-of-dried-human-tears-180947766/.

Team, Family Health. "Why Do We Cry? the Truth behind Your Tears." Cleveland Clinic. Cleveland Clinic, February 21, 2022. https://health.clevelandclinic.org/tears-why-we-cry-and-more-infographic/.

Zoffness, Rachel, and Karen Schader. *The Chronic Pain and Illness Workbook for Teens: CBT and Mindfulness-Based Practices to Turn the Volume down on Pain.* Oakland, CA: Instant Help Books, 2019.

Zoffness, Rachel. *Pain Management Workbook: Powerful CBT and Mindfulness Skills to Take Control of Pain and... Reclaim Your Life.* S.l.: READHOWYOUWANT, 2021.

Made in United States
Troutdale, OR
02/20/2024